A History of Irish Farming, 1750–1950

A History of Irish Farming
1750–1950

JONATHAN BELL
&
MERVYN WATSON

FOUR COURTS PRESS

Set in 11 pt on 14.5 pt Garamond by
Carrigboy Typesetting Services, for
FOUR COURTS PRESS LTD
7 Malpas Street, Dublin 8, Ireland
e-mail: info@fourcourtspress.ie
http://www.fourcourtspress.ie
and in North America for
FOUR COURTS PRESS
c/o ISBS, 920 NE 58th Avenue, Suite 300, Portland, OR 97213.

ISBN 978–84682–208–7 pbk

SPECIAL ACKNOWLEDGMENT

The financial support of Bord Bia in the production
of this work is gratefully acknowledged

Irish Food Board

Printed in England
by MPG Books, Bodmin, Cornwall.

Foreword

IRISH PEOPLE HAVE an obsession with owning property whether it is in the form of land or houses or both. In reading this book one has a clearer understanding of the genesis of this obsession. While ownership of a very small plot of ground in which to grow potatoes was no guarantee of survival it was viewed by many as their best way of reducing their vulnerability to starvation and disease. The disagreement among the commentators of the day regarding the desirability or otherwise of labourers owning land for supply of food for their families exemplifies the huge social divide between the residents of the 'big house' and the occupiers of the 'bothóg' or the 'scalp'.

Bell and Watson's encyclopaedic knowledge of the houses and farms, the implements and machinery, the crops and the farm animals is evident as is their passion for the subject. The authors don't just focus on the technical aspects of farming and its development but by giving us the views of individual farmers, labourers and commentators they also bring to life for us the story of the development of rural society in Ireland. The style engages the reader as if part of a story-telling evening making the experience thoroughly enjoyable.

We consider farming methods today to be complex and sophisticated mainly as a result of mechanization, however, when we read this book we appreciate that, before the era of the tractor, the business of farming was indeed complex albeit that the tools were basic and the labour manual. The development of land where none existed, the refinement of the spade to exploit local conditions and the interest in selection of better quality breeding stock are all good examples of a people not only needing to survive but also wanting to improve and pass on a better way of farming to the next generation. Minimal cultivation, as in the case of the lazy-bed techniques, condemned by many as indicative of an unwillingness to work, has been shown to have a valid place in modern farming practices. In its day it was a particularly innovative method of bringing marginal land into cultivation.

This book will cause readers to pause and think about how farming has developed in Ireland and how the methods and systems that we take for granted today arose from a combination of necessity and ingenuity.

Michael Maloney
BORD BÍA

Contents

Illustrations

FIGURES

PLATES

appear between pages 192 and 193

Introduction

THIS BOOK SUMMARIZES the findings of thirty years' work in the Ulster Folk and Transport Museum. During that period, both authors worked at collecting and interpreting material related to the history of Irish farming. This led to a number of publications, and the book attempts to draw on the main findings of these pieces of work, and put them together in one place, along with previously unpublished material.

The book concentrates on the practicalities of farming; implements and techniques, crops and livestock. The dates given in the title are to some extent arbitrary. Some of the farming methods described have been used for centuries, and a remarkable number are still practised. Describing these aspects of farming has sometimes meant going back in time before 1750, or up to the present day.

Many of the farming methods we have documented were recorded during fieldwork. This was mostly carried out in the nine historic counties of Ulster, and related to small farming life, where local ingenuity and refinement was most obvious. However, we found that these findings often had relevance to other parts of Ireland, and we hope that any northern bias in the data will be understood in this context.

The work of investigating Irish rural life has many joys, including the opportunity to travel around Ireland and find out more about the economic and social influences that produced the beautiful country we live in. The greatest joy, however, came with meeting many people who have become long-term friends. All the people who patiently explained local aspects of farming deserve thanks. They include Malachy McSparran, Bertie Hanna, John and Maeve O'Connor, Joe Kane, Dolly McRoberts, Anthony Brennan, Fred Coll, Hugh Paddy Óg Ward, and Joe Kane. Friends and colleagues from other institutions also helped with the production of the book. Thanks go to Michael Maloney, Martin Fanning, Liam Corry, Toddy Doyle, Philip Flanagan, Brendan Dunford, Finbar McCormick, Séamas MacPhilib, and Robbie Hannan. We are also grateful for the institutional help we received from The National Museum of Ireland, the Ulster Folk and Transport Museum, the Ulster American Folk Park, and Muckross House Traditional Farms in Killarney. Special thanks go to Bord Bia for the generous sponsorship that allowed the inclusion of the book's colour images.

CHAPTER ONE

Rural Society and Farming Methods

THE FIRST FARMERS arrived in Ireland more than 6,000 years ago and evidence for agricultural settlements and enclosure has been found from that period onwards.[1] Prehistoric field enclosures have been uncovered at sites such as the Ceide Fields in County Mayo, and there is ongoing, if sporadic, evidence for enclosures in the Early Medieval and Post-Medieval landscape. However, most of the landscape of fields enclosed with hedges and walls that we see today is relatively modern, dating from the eighteenth and nineteenth centuries, the period covered by this book (fig. 1). In this chapter we will look at some aspects of the social structure and associated farming systems that produced this landscape.

LANDLORDS AND THEIR ESTATES

Between 1700 and 1900, agricultural land in Ireland was controlled by landlords. By the 1780s, 5000 landlord families owned over 95% of all productive land in Ireland, and this number rose to almost 10,000 by the mid-nineteenth century. Landlords' estates could be vast. Three hundred landlords owned estates of more than 10,000 acres, and 3,400 owned estates of between 1000 and 10,000 acres.[2] These people could be expected to have major impacts on the landscape. However, this was not often the case.

At the beginning of the eighteenth century, the vast majority of landlords were recent incomers of English or Scottish origin. Throughout most of Ireland, the new landowners were Protestant and English speaking, while their tenants were Irish speaking Catholics. By 1703, only 14% of land in Ireland was owned by Catholics, and this figure was reduced even further during the eighteenth century by the implementation of Penal Laws, designed to prevent Catholics from inheriting undivided estates. Landlords and tenants often saw themselves as having a separate ethnicity and culture.

There has been an ongoing debate in Irish historical writings about the role of landlords in Irish society, and in the development of agriculture. 'Revisionist' historians have argued against what they see as an entrenched nationalist view that landlords were ruthless exploiters of the poor, who contributed nothing

1 Mourne Mountains, Newcastle, County Down, *c*.1910 (Lawrence Coll. *c*.1652).

to the wellbeing of their tenants, and there is some evidence supporting this view. Some landlords did engage in large-scale developments on their estates, and a number of nineteenth-century landlords made radical attempts to rationalize farms on their land, for example, by abolishing the rundale system, which will be discussed below. However, the best quantitative evidence suggests that in general, landlords put relatively little money into developing their estates. Cormac Ó Gráda has estimated that overall, they invested as little as 3% of their wealth in agriculture.

> Why did Irish landlords do so little? In part because they, unlike English landlords, were wholesalers rather than retailers in land, and therefore cared or knew little about the circumstances of individual farms ... In the late eighteenth century, most tenants still rented from middlemen rather than directly from landlords.[3]

2 Steam ploughing demonstration, 1860 (G.E. Fussell).

Despite important exceptions, landlords in general were not central to the development of Irish farming. Ó Gráda has pointed out that in terms of efficiency, economic theory suggests that there is little to choose between a system of rent paying tenants and one of peasant proprietorship.[4] During the 1880s there was a struggle for ownership of land, which has become known as the Land War. By 1900, the landlord system was disappearing. 'What is striking … is how easily the [landholding] system imposed by the sword in the seventeenth century was eliminated in the nineteenth and early twentieth centuries and how few traces it left.'[5]

However, if landlords as a class did not fulfill their role as developers and improvers, we will be referring to the work of a minority who did contribute to the introduction and spread of many of the farming methods described in this book. The estate farms of progressive landlords provide some of our best evidence of risky experimentation, either with large-scale machinery such as steam ploughing equipment (fig. 2), or the adoption of new ways of utilizing older techniques such as working with oxen (fig. 3). Some landlords were also involved with the establishment and the work of Farming Societies, and their publications. The correspondence columns of publications such as the *Irish Farmers Gazette*, or the *Munster Farmer's Magazine* give us some of our clearest contemporary evidence for ideas about techniques and systems and the discussions they provoked.

3 Ploughing with Kerry oxen, Sir Eustace Beacher's estate, Ballygibbon,
County Cork, *c.*1916, photo courtesy Bob Peters.

TENANT FARMERS

For most of the period covered by this book, the key relationships for farming
methods were those between wealthy farmers and smallholders or labourers.
Differences in the scale of farming practised by tenant farmers could be
extreme. In North Tipperary, for example, Arthur Young visited a farmer, a Mr
McCarthy, a tenant farmer whose farm of 9,000 acres at Springhouse, Young
believed 'must be the most considerable one in the world.' The livestock kept
on the farm give some idea of the scale of production.

8,000 sheep	200 two-year olds
2,000 lambs	200 three-year olds
550 bullocks	80 plough bullocks
80 fat cows	180 horses

Mr McCarthy also employed between 150 and 200 labourers.[6]

Huge farms like this were generally found in the wealthiest farming areas in
Ireland. Kevin Whelan has identified three of these: the pastoral region,
covering much of North East Leinster and East Connacht; the dairying region
of Kilkenny, West limerick and North Kerry; and 'the tillage triangle,' linking
Cork, Louth and Wexford.[7] As with the estate farms managed by progressive

Prize Plough

AGRICULTURAL IMPLEMENT
AND
MACHINE MANUFACTORY,
53, 55 AND 57, MAY STREET, BELFAST
GRAY & CO.
(*Late GRAY & SON*),
Beg to acquaint their numerous Friends and Customers that
they continue to manufacture
IMPLEMENTS AND MACHINERY,
Of every description at the above address, and that all Orders
now entrusted to their care will have the strictest attention.
Their long practical experience, the numerous flattering
Testimonials they have received from Noblemen and Farmers,
and the various Medals they have received at the Shows they
have attended, warrants them in stating that no other house in
town can supply Implements or Machinery of the same quality,
to suit all kinds of land, on such reasonable terms, and thereby
anticipate an extension of the already large share of patronage
bestowed on them.
Price Lists, with Photographic Views of the different Implements, may be had on application at the Factory.
Repairs of every description promptly executed in the most
satisfactory manner.
*NOTICE.—Gentlemen and Farmers sending their Servants with
Implements to be Repaired, should be careful to instruct them where to go, as
mistakes sometimes occur,*
GRAY & CO.
5 55 and 57, MAY STREET, BELFAST.
The only Manufacturer of the Prize Implements.
NO CONNEXION WITH ANY OTHER CONCERN
Freight paid to Dublin, Londonderry and Liverpool.

4 Advertisement for Gray's
foundry, Belfast, *c.*1848.

landlords, the big tenant farms in these areas provide evidence for risky investments in new equipment. At the level of medium-sized tenant farms in these regions and elsewhere, we find the earliest evidence for the introduction of new, standardized farm machinery, such as all-metal ploughs, reaping machines or horse drawn potato diggers (fig. 4). These medium-sized commercial farms were also in the forefront of the development and introduction of standardized breeds of farm livestock during the later nineteenth century.

In the much of the west of Ireland, and especially on marginal land, many tenant farms were very small, and provision of a subsistence living for the farming family was a major goal. It is on farms such as these that we find the biggest reliance on manual labour, and also evidence for implements such as spades, sickles or flails, used in techniques which often showed great refinement in their adjustment to local conditions (fig. 5).

One region of small farms, described by Kevin Whelan as a proto-industrial zone,[8] mostly concentrated in Ulster, was the result of the spread of commercial activity, the home production of linen. In 1780, Arthur Young was appalled by the effects of linen production on farming in this region.

5 (*left*) Anthony Brennan of
Aughavas, County Leitrim
using a loy spade in 2004.

6 Beetling, scutching and
hackling flax, County Antrim,
c.1780 (W. Hincks).

> There … is not a farmer in a hundred miles of the linen counties of
> Ireland. The lands are infinitely subdivided, no weaver thinks of
> supporting himself [only] by his loom; he has always a piece of potatoes,
> a piece of oats, a patch of flax, and grass or weeds for a cow … Ten acres
> are an uncommon quantity … four, five, or six, the common extent.[9]

The amount of labour involved in processing the flax meant that most weaver-
farmers could only work very small holdings (fig. 6). This led to the spread of
tiny farms even in areas of fairly good land, and unlike proprietors in
commercial farming areas of Leinster and Munster, Ulster landlords
encouraged the spread of tiny holdings. The industrialization of the linen
industry in the nineteenth century, and a growing dependence on imported
flax, almost wiped out the weaver-farmer economy, but holdings in Ulster
remained small, even by Irish standards. In 1911, almost 90% of Ulster farms
were less than 50 acres, as compared to 84% in Ireland as a whole.[10]

LABOURERS

Agricultural labourers became important in Ireland during the eighteenth
century. Relationships between affluent tenant farmers and smallholders or
labourers could be tense. They can often be understood in terms of the
potential conflict between dominant commercial interests and a struggle for
economic and social survival. The conflicts between 'improved' and 'common'
farming practices were also often an expression of these differing priorities.

It is not easy to identify different types of workers, because even the
distinction between farmers and labourers was often unclear. The main
difficulty is in defining people as either landholders, or landless. Many people
who spent most of their time labouring for others still described themselves
as farmers, if they had access to even a small patch of land. Throughout the
nineteenth century, commentators recognized this as a question of status. In
1887, for example, the British Registrar General explained, 'They will not
demean themselves by calling themselves labourers.'[11] Many labourers
negotiated ways in which to secure a small area of land, at least for one
growing season, not so much for status, as to attempt to secure food for their
families.

CONACRE

Contracts where workers are hired for money wages paid by the day or the week are now standard aspects of working life in rural society. In the eighteenth century, however, these wage earners were mainly found in the north and east of Ireland, where money was most frequently used, and even here access to land for cultivation was still very important. Day labourers often used their cash earnings to purchase 'conacre', a patch of land rented for a season, where they could raise a crop – especially potatoes – on which their family could subsist. The patch of land could vary between one eighth of an acre, to two acres (0.13 to 0.8 hectares).[12]

Farmers made big profits from conacre. In 1829, Joseph Lambert estimated that if a farmer rented 200 acres from a landlord, and let out 150 acres of it in conacre, he could expect to make a clear profit of £2,500 within four years. Conacre, Lambert claimed, was 'the system which yields most profit in Ireland, within existing conditions'.[13] The labourers were willing to pay these rents because they could generally provide themselves with potatoes below the market price, as well as procure waste fodder with which to fatten a pig. Conacre also allowed labouring families to utilize the labour of women and children who otherwise would be underemployed. In the nineteenth century, conacre was almost universal among wage labourers in County Cavan, for example, but the high rental cost meant that many workers often had to travel for seasonal work in England or Scotland in order to pay for it.[14]

Conacre was a strategy adopted mostly by labourers who were paid by the day or the week. Male and female servants who lived on the farms of the people who employed them were paid mostly in kind rather than cash. Their contracts would often include food and lodgings. Some of these workers were also given access to a small piece of land. For many workers, who became 'cottiers' or 'cottars', the exchange of labour in return for a small patch of cultivable ground and a dwelling house was the defining aspect of the working relationship.

COTTIERS

In the decades preceding the Famine, 'cottars' or 'cottiers' provided a lot of the manpower required for labour-intensive tillage farming. Historians have suggested that the cottier system was a response to underemployment and the lack of a fully developed money economy. By taking men on as cottiers, farmers avoided having to pay for labour in cash, while many labourers valued the security of a cottage and potato ground. Arthur Young summarized how the system operated in the 1770s. He believed that the cottier system of labour in Ireland,

> was probably the same all over Europe before arts and commerce changed the face of it. If there are cabbins on a farm they are the residence of the cottars, if there are none, the farmer marks out the potatoe garden, and the labourers … raise their own cabbins … A verbal contract is then made, that the cottar shall have his potatoe garden at such a rate, and one or two cows kept [for] him at the price of the neighbourhood, he finding the cows. He then works for the farmer at the rate of the place … a tally being kept … and a notch cut for every day's labour; at the end of six months, or a year, they reckon, and the balance is paid. The cottar works for himself as the potatoes require.[15]

Cottiers were anxious to be kept on for more than one year, and this made them very dependent on their employers. Tenant farmers who employed cottiers could impose various restrictions on their activities. For example in the decades before the Great Famine, tenant farmers in County Cavan prevented their cottiers from keeping chickens, to reduce competition in the growing egg trade.[16]

Where a house and potato ground were all that was provided, the arrangement was sometimes known as a 'dry-cot'. An agreement to provide grazing for a cow as well as potato ground was known as a 'wet-cot'. Turf cutting rights were sometimes included in the deal. However, Arthur Young's description of the system suggests that even in the eighteenth century, work and goods or land given in return were calculated by some reference to equivalent monetary rates, and in the early nineteenth century, monetary transactions continued to replace payments in kind. The cottier system almost disappeared with the Great Famine of the 1840s. Many cottiers died from disease or starvation, or survived only by emigration. Most of those cottiers who stayed in rural areas of Ireland after the Famine became labourers depending on a money wage.

SUBDIVISION

The hazy line dividing farmers and labourers becomes very clear when we look at landholding patterns among the poorest sections of the rural population, and especially in the west of Ireland. People wanted access to land to secure a

7 Fields on the Aran islands, County Galway. Subdivision here was not only due to population pressure. The walls were built from stones cleared from fields, and gave shelter from Atlantic gales.

living for themselves and their families, and as population rose from the mid-eighteenth century, this desire often gave rise to ongoing subdivision of farms, particularly in these poorer parts of Ireland, as farmers divided their holdings between their children. By the early nineteenth century, chronic subdivision had produced patterns of landholding that were seen as absurd, with farmers sometimes cultivating dozens of tiny patches that put together amounted to no more than a few acres of land (fig. 7).

Subdivision was most common on estates where landlords were particularly lax in estate management. In the richest parts of Ireland, especially in the lowlands of Leinster and North Munster, commercial farmers resisted attempts at subdivision, and tiny holdings were mostly confined to hilly areas of marginal land.

RUNDALE

One strategy adopted by country people to get access to land, was to participate in the rundale system of landholding. The system was one of the most common types of settlement on marginal land during the eighteenth and nineteenth centuries. There is ongoing discussion between scholars as to how old and how widespread rundale was, but by 1750, it was dominant in large areas, especially of poor, marginal land in the north and west of Ireland (fig. 8).

8 A 'fossilised' rundale infield at Ballyliffin, County Donegal, 1960s.

Rundale settlements usually had a number of distinctive characteristics:

- a cluster of houses and outbuildings (the *baile*, 'town' or 'clachan')[17] formed the core of the settlement. This usually consisted only of dwellings, outbuildings and tiny gardens. Buildings such as churches, shops or schools were not usually found in the settlement.
- complex kinship ties linked many of the families living and working in the settlement. The land around the clachan was worked by the families living there. This was known as the *infield*. Each farmer's holding was made up of unenclosed, scattered plots of good, medium and poorer quality land. The arable strips were unfenced, but were marked out in a number of ways, most often by small earth banks known in different areas as *mearings, ribs, roddens, bones, rowins*, or *keelogues* (fig. 9)[18]
- A larger area around the *infield* was used for grazing, and occasional cultivation. This was known as the *outfield*.
- Mountain or other marginal land in or near the settlement was used as common grazing, usually during the growing season, when crops in the unenclosed strips in the infield were vulnerable to straying animals. The seasonal movement of animals to this common land was known as booleying (*buailteachas*).

9 Mearings in the disused infield at Machaire Gathlán, County Donegal.

Many settlements had a leader, known by different names, including *maor, rí,* 'king' or 'herd'. This man negotiated with the landlord or his agent on behalf of the other people in what was in fact a joint tenancy.[19]

The egalitarian drive behind the rundale system was especially clear when the participating farmers practised 'changedale.' Here, instead of fixed allocations of ground within the infield and outfield, plots were reallocated periodically. In the 1770s, Arthur Young noted the practice in several places, 'which is every man changing his land every year'.[20]

Close ties bound people living in clachans, especially when the rundale system was fully operational. Despite landlord complaints about the conflicts that erupted over issues such as distribution of arable strips, or straying livestock, people valued the communal life of the clachans. As late as 1908, one commentator said that on Rathlin Island in County Antrim, 'holdings were held in rundale ... Redistribution of land was very desirable, but redistribution of people was a great difficulty, as they did not wish to leave their homes.'[21]

The values underlying the rundale system have been summarized by James Anderson. '[Rundale] was based more on communal than on individual enterprise, originally in kinship groups, later on partnership farms. Co-operation and equity were among its guiding principles, though by the nineteenth century … more competitive and individualistic attitudes often prevailed.'[22]

The evidence for rundale in areas such as the Glens of Antrim supports Desmond McCourt's view of the system as it operated in Ireland since 1750. [23]

> Often … in comparing estate maps of the seventeenth and eighteenth centuries with the ordnance survey maps of the nineteenth century, we seem to be dealing with settlements in which the nucleated condition was not original, but consequent on an initial single-family settlement, which, with increased population, swarmed eventually to form a wider kin group, or 'clachan'. On the other hand, the reverse process – a secondary dispersal of houses following the break-up of the open-field system and the dissolution of the 'clachans' – was equally endemic and particularly rapid in the nineteenth century. Thus things were in a constant state of flux … All this would suggest that we are dealing, not with opposing types of settlement, but with a single dynamic scheme, within which dispersal and nucleation were alternative developments.[24]

The co-existence of clachans and dispersed settlement can be clearly seen in the Glens of Antrim. First edition ordnance survey maps made of the area in the early 1830s, often show both side by side.

It is important to bear in mind the flexibility of arrangements that could be described as 'rundale' or booleying'. For as long as detailed records exist, it is clear that Irish farmers mixed old and new techniques, and general farming systems, in ways that suited their own situation. Attempting to impose rigid patterns on data relating to settlement can therefore distort history, by ignoring the intelligence and adaptability of the people who created these arrangements. Rundale may sometimes have been adopted as a solution to subdivision. A rundale arrangement could have been put in place after the subdivision of holdings over several generations led to the need for a communal system for making decisions, which could ensure viable planning of cultivation and grazing.

Subdivision of holdings was not an inevitable consequence of farming using rundale, but the two were often associated. In most nineteenth-century rundale settlements that we know of, changedale did not operate, so families might keep the same strips of land for more than one generation. If strips were subdivided between heirs, to ensure that they had land on which to grow a subsistence crop of potatoes, fragmentation of holdings could be extreme. In one case in Donegal in the 1840s, for example, 'one man (a tailor by trade) had his land in 42 different places and gave it up in despair, declaring that it would take a very keen man to find it'.[25] The 'disintegrating process' could be remarkably rapid. Desmond McCourt cites another example from County Donegal, where within two generations, a single farm of 205 acres had fragmented into 29 holdings scattered in 422 plots.'[26]

THE END OF RUNDALE

The rundale system seems to have survived longest in remote places like Rathlin Island off the coast of County Antrim. In 1908, a Government Commission found that during the Revd Gage's time as landlord, 'rundale was universal on the island … Mr Gage had not tried to do away with it', and that this was still the case in 1908, when it was claimed that 'nearly the whole island was held in rundale'. (Another witness stated that *some* holdings were held in rundale.)[27]

Recent fieldwork on the island suggests that elements of rundale farming have survived on Rathlin much longer than even this late date, until well within living memory. In 2006, an islander, Mr Bertie Currie, told the authors that he had worked for two farmers in Brockley townland on the island during the 1970s. He said that they operated a 'kind of rundale system'. Bertie was not always able to tell whose land he was working on. Some of the field boundaries were no more than low earth banks. He said that the two farmers enjoyed his confusion, and would give him 'a clip on the ear' when he made a mistake. The lack of emphasis on enclosure, which was associated with the rundale system, had particularly drastic results on cliff-bound Rathlin. In 1907, it was claimed that cattle, sheep and horses had fallen off the cliffs, because they liked to graze the sweeter grass near the edges. Apart from saving these poor animals, one witness pointed out that 'Fencing would allow the children to go to school, which they could not do now, as they had to watch the cattle.'[28]

Many landlords condemned rundale as inefficient, and a cause of disputes between tenants. Lack of enclosure between strips was said to cause conflict, especially when livestock wandered from a strip that was left for fallow grazing, into a neighbour's crops. The lack of enclosure on many Irish farms had been noted in the late eighteenth century. In 1780, Arthur Young was amused by the extent to which Irish farmers hobbled their livestock to prevent them straying.

Many of the cocks, hens, turkies and geese, have their legs tied together to prevent them from trespassing on the farmer's grounds. Indeed all the livestock of the poor man in Ireland is in this sort of thralldom; the horses are all hopping about, the pigs have a rope of straw from around their necks to their hind legs.[29]

The process of allocating arable strips was said to be a particular source of conflict, because of disputes over who got what. The periodic redistribution of the strips was also claimed to discourage individual landholders from attempting long-term improvements to the quality of land. In many areas, landlords were attempting to abolish rundale during the 1830s and 1840s. As with landlord attempts to encourage the colonization of waste or marginal land, the re-organization could be more or less radical. On the Londonderry estate in the southern part of the Glens of Antrim, for example, the system was abolished as leases fell in. A report on the estates of the marquis and marchioness of Londonderry made in 1844 by the land agent John Lanktree, includes the following.

> The tenants in … Galboly I have also viewed for an increase [in rents] but have not received the map of it from the surveyor. In … the above [case] … I hope to affect a permanent improvement of the lands by abolishing the pernicious farming called rundale.[30]

In 1845, Lanktree reported that rundale had been abolished at Garron Point, in a way which he saw as considerate to the tenants.

> [At] Garron Point … [which was] farmed promiscuously in Rundale and covered with a batch of small dirty and unhealthy cabins – a great improvement has taken place. Six new houses of excellent workmanship have been erected – new fences constructed, rundale abolished and the tenants vie with one another in their most praiseworthy efforts. The clearance system would have doomed each happy cottager on this Cape of Storms to wandering, indigence or premature destruction.[31]

In areas where rundale was widespread, modern field patterns often reflect landlord interventions, aimed at abolishing the system, and rationalizing farm size and layout. In many parts of the north and west of Ireland, landlords attempted to replace rundale by providing tenants with a range of land quality by laying out long narrow farms, running from lowland to moorland. This created the 'ladder pattern' of fields, typical of many areas where land runs up into surrounding hills. The process of laying out these long, narrow holdings was known as 'striping'. Ladder farms which extended from valley bottoms to hilly margins preserved the rundale principle of shares in both good land and bad in holdings.[32] Striping of holdings generally began in the west of Ireland in the early nineteenth century, and was continued by the Government funded Congested Districts Board, into the twentieth century.

One of the best known instances of an estate energetically 'improved' by a landlord, was the project undertaken at Gweedore by Lord George Hill. Hill's account of his work[33] was widely discussed both by defenders of the landlord system in Ireland, and reformers who condemned it. Hill saw the abolition of the rundale system as his main achievement. He was anxious to show the good effects of all aspects of his project, however, and he did this by contrasting the way of life of tenants on the Gweedore estate when he took it over, with the situation at the end of his proprietorship. He emphasized the destitution and backwardness of the people on the estate. Hill claimed that the way of life in Gweedore was essentially 'nomadic', tenants moving from the mountains to the coast and offshore islands in search of grazing for cattle at different seasons of the year.[34] This transhumance had been documented in Donegal since at least the early seventeenth century, although it may be that by the time Hill was writing the pattern of seasonal movement was less systematic and unchanging than he made it appear. Oral evidence and documentary records from Gweedore suggest that seasonal movements of people and livestock did occur, however, people often living in the hills for some period of the year, usually in a temporary dwelling known as a *bothóg,* which will be described in the next chapter.[35] Hill was certainly right to emphasize the importance of cattle for meat, milk, and dung. The collection of dung for manure in byre-dwellings was claimed to be common:

> Man and beast [are] housed together, i.e. the families at one end of the house, and the cattle in the other end of the kitchen. Some houses having within their walls, from one cwt. to 30 cwts. of dung, others having from ten to fifteen tons of weight of dung and are only cleaned out once a year![36]

10 *Gealasacha Dhoir*, the 'Braces of Gweedore'.

Hill's account provides an excellent case study of the different ways in which 'improving' activities were perceived by a landlord and tenants. Apart from conflicts over the abolition of rundale, Hill caused great resentment when he reserved 12,000 acres of mountain grazing, nearly half the area of the Gweedore estate, for his own use. In 1855, he let portions of the mountain to Scottish graziers who imported large numbers of sheep. Hill's celebrated campaign against rundale in Gweedore also involved the creation of strip farms of such narrowness and length, that they are popularly known as *Gealasacha Dhoir*, the 'Braces of Gweedore' (fig. 10).[37] Some of the new holdings were a mile long and so narrow, sometimes as little as seven or eight feet wide, that new houses built on them had their narrow gable ends facing the road, rather than their frontage.[38]

CONSOLIDATION

Just before the Famine of the 1840s, two-thirds of the Irish population depended on farming for their livelihood. There were 685,309 farms, well over three times the number of holdings today. Forty-five per cent of farms were between one and five acres, and only 7% were over 50 acres.[39] Since the mid-nineteenth century, the average size of Irish farms has increased. In part this was due to emigration from the countryside. Hilly areas especially became depopulated, and the small farms established on them were no longer viable. All over Ireland the implementation of the land acts of the late nineteenth and early twentieth centuries produced owner-occupied farms that proved too small to be economically viable, and their owners sold up. This allowed enterprising farmers to rent, or buy the land which then became available, and increase the size of their own holdings.

Between the 1840s and 1911, the number of holdings between one and five acres fell from 182,000 to 62,000, and the average farm size was 15–30 acres. However, 200,000 holdings, supporting about one million people remained, mostly on poor land. These tiny holdings were not viable.[40] By the end of the twentieth century, the average size of farm in Ireland was 30 hectares (almost 75 acres), and it has been predicted that the number of farmers will continue to decline dramatically.[41] Consolidation not only removed distinctive settlement features associated with rundale, or the cottier system, it also led to the spread of standardization and specialization that first appeared among the medium-sized tenant farms described above. Farming systems, implements, crops and livestock became less regionally distinctive and the farming economy became increasingly tied in to international markets. The rest of this book will attempt to describe these processes in more detail.

CHAPTER TWO

Farm Houses and Farm Yards

Throughout the last three centuries, rural housing has reflected the extremes of wealth and poverty in the Irish countryside. At the upper end of the social scale, eighteenth-century landlords lived in the 'Big Houses' that are now admired as one of Ireland's important contributions to Europe's architectural heritage. Less grand, but still well-off, minor gentry of the same period also lived in fine houses, with two or three storeys, and slated roofs. However, Ireland did not have a large number of these well-off farmers. Wealthy squires and comfortably-off yeomen were much more numerous in England. In Ireland, even in the affluent south-eastern counties, there were relatively few people living at levels between the lifestyles enjoyed by the very rich, and those endured by the bulk of the rural population. The houses of the latter, the mass of small farmers and labourers, and their associated farm buildings, will be the main subject of this chapter.

FARM HOUSES

The earliest surviving remains of small rural houses in Ireland date from far back in prehistory, but very few houses lived in by ordinary people survive intact from earlier than 1700. We know from archaeological excavations that rectangular houses with a door in the long wall had become fairly common in Ireland by medieval times, and illustrated maps from around 1600 show a number of types of dwelling, including single-storey thatched houses, again with a door in the long front wall. Cottages that survive from 1700 onwards, built by small farmers, either for themselves or their labourers, also show these features as elements of well-developed 'vernacular' designs. The houses are usually single-storey, with between one and three rooms that open directly into one another without any passage or central hall (fig. 11).

Two basic types of vernacular house have been identified, each of which was associated with particular areas within Ireland.[1] The main feature used in classifying the houses is the relationship between the kitchen hearth and the front door. 'Direct entry' houses have a kitchen with a hearth built in to one end wall, and the entrance door in a side wall at other end (fig. 12). 'Hearth-

11 House at Kilmore Quay, County Wexford, *c.*1965, photo courtesy
National Museum of Ireland.

lobby' houses have a front entrance facing side-on to the kitchen hearth, the
door and hearth being separated by a small 'jamb' wall (fig. 13).

Direct-entry houses are mostly found in the north and west of Ireland, and
hearth-lobby houses in the southeast and midlands. However, there are a lot
of variations within this basic classification. For example, some vernacular
houses are two storeys in height, and their gable ends can be hipped or
straight. Many houses also have a greater number of internal divisions, creating
more rooms and passages.

Most of these surviving rural houses are thatched, or were originally
thatched. Fred Aalen has summarized the basic characteristics of their roof
construction, including the different materials used.

> In most areas a thin layer of sods or 'scraws' is laid upon and tied to the
> roof timbers to support the thatch and improve heat insulation.
> Wheaten straw is the most widely used for thatch. Northern Leinster,
> however, preferred oaten straw, and reeds were favoured in some
> localities in Munster and flax in counties Londonderry, Donegal and
> Fermanagh. Rushes and tough marram grass were used in mountainous

coastal areas in the northwest and west. The most general method of securing the thatch is by pinning the straw to the scraws with scollops – pegs made with thin rods of briar and hazel. Along the northern and western coasts, presumably in response to the strong winds, the thatched roof is commonly held in place by a rope tied to pegs in the house walls or to a row of stone weights.[2]

Despite variations in lay-out and construction, in the past many vernacular houses had very similar living arrangements. The two important day-time spaces were usually the kitchen, and what was known as 'the west room,' or simply as 'the Room'.

A description of how these spaces were used, recorded in County Clare in the 1930s, shows the differences between them quite clearly.

> The central room … was the kitchen. There the family passed their lives and prepared and ate the food cooked in the hearth … behind this hearth was another room … it was called for its position in the house, the 'west room' … [This] room was a sort of parlour into which none but the most distinguished guests were admitted. In it were kept pictures of the dead and emigrated members of the family, all 'fine' pieces of furniture.[3]

Many vernacular houses continued in use into the late twentieth century. Henry Glassie's lyrical description of life styles in the Ballymenone area of County Fermanagh during the 1970s shows that here also the main room in the house was the kitchen, with several other less-used rooms flanking it. His description of the kitchen, and the lay-out of the house in general, emphasizes how the structure and layout of the houses reflected the communal lifestyle of the people who lived in them.

> Knowing that all the houses have the same plan … people gain a sense of community … The houses present similar facades, one-storey and sparkling white … Noting the door's location, in relation to its windows, one for each room, the visitor knows the house's plan … The kitchen is not a private place. It is gracefully vulnerable, open and occupied. People from without pierce quickly to its very center, where a fire burns, ready to boil water for tea.[4]

12 Direct entry house form Magilligan, County Derry, rebuilt in the Ulster Folk and Transport Museum.

13 Hearth lobby house from Corradreenan, County Fermanagh, rebuilt in the Ulster Folk and Transport Museum.

HOUSING AND POVERTY

The houses just described are now seen as typical vernacular rural dwellings and preservation of the small number that survive has rightly become a priority for conservationists. However, most contemporaries who wrote about rural housing during the last three centuries were appalled at the terrible living conditions endured by masses of Irish country people. The following section will describe some of the houses of the rural poor, and attempts to improve them.

Arthur Young's description of a typical 'cabin' of the 1770s (fig. 14) was echoed in many later accounts.

> The cottages of the Irish, which are called cabins, are the most miserable looking hovels that can well be conceived: they generally consist of one room … and have only a door which lets in light instead of a window, and should let smoak out instead of a chimney, but they had rather keep it in … the smoak warms them, but certainly is as injurious to their eyes as it is to the complexions of the women, which in general in the cabins of Ireland has a near resemblance to that of a smoaked ham … The bad repair … [the] roofs are kept in, a hole in the thatch often mended with turf, and weeds sprouting from every part, gives them the appearance of a dunghill.[5]

14 Cottage at Sheskinbeg, County Donegal, *c.*1880 (Glass Collection, UFTM).

Most travellers recognized that housing conditions varied by region within Ireland, but in the early nineteenth century, and especially around the period of the Great Famine of the 1840s, observers were increasingly horrified by the conditions they recorded. In 1842, William Makepeace Thackeray was particularly shocked by hovels which he saw at Bantry, County Cork. He claimed that a 'Hottentot kraal' would be a more comfortable dwelling. A French traveller of the same period claimed that the ultimate depths of human misery could be found in Ireland, 'worse than the negro in his chains.'[6]

The most notorious form of dwelling was the 'scalp' (Irish: *scailp*) (fig. 15). Scalps hardly qualified as buildings. They were very crude structures built up in ditches, or under rocks, using branches or sods to provide extra shelter. Arthur Young described how they were set up.

> A wandering family will fix themselves under a dry bank, and with a few sticks, furze, fern, etc. make it into a hovel much worse than an English pigstie … In my rides about Mitchelstown, I have passed places in the road one day, without any appearance of a habitation, and next morning found a hovel, filled with a man and woman, six or eight children, and a pig.[7]

Scalps were often the last resort of homeless, landless people, many of whom were obliterated in the Great Famine of the 1840s. A slightly more substantial type of dwelling was known as a *bothóg* (fig. 16).

This was usually built into the side of a hill, so that the cut-away slope formed the back and part of the side walls. The front wall and rest of the sides were built up using stones or sods. Bog timber was used to make a purlin roof, which was then covered with strips of turf, or thatched. Smoke from the hearth, which was set into the back wall, escaped through a hole in the roof. *Bothógaí* were often built for summer use by people tending cattle in upland pastures during summer – booleying – but they were also used as permanent dwellings by very poor people.

In the Gweedore area of County Donegal as late as the 1980s, a family connection with a *bothóg* could be used as a taunt. 'Sure, what would you know about it? Your grandfather lived in a *bothóg*!' However, an incident remembered in the same part of Donegal, also shows the ability of people to find joy even in conditions of terrible poverty. In this account, the construction of a *bothóg* was an occasion for celebration. A young couple from Gola island went to the Gweedore mainland to be married. When they left the

15 Scalp in Connemara, in the 1840s, from Hall's *Ireland*.

16 *Bothóg* Mhary Mhóire in Gweedore, *c.*1880 (Glass Collection UFTM).

17 Fred Coll's *bothóg*, Doire Beaga, County Donegal, built for a
BBC television programme in 1995.

island, their neighbours began to build a *bothóg* in which they could set up
home. The newly-weds arrived back on the island that evening to find that
their home had been built, a pot of potatoes was boiling on the fire, and a
dance was held in the *bothóg* to celebrate the marriage.[8]

Some people in the west of Ireland still remember making and living in
bothógaí (fig. 17). The structures received a last lease of life during the Second
World War. The Irish Free State remained neutral in the war, but severe
shortages of food and fuel were common. One of the ways in which the Irish
government tried to deal with the fuel situation was to organize massive turf
cutting schemes, where work continued even after dark by electric light.
Workers sometimes found it easier to sleep out on the bog than to travel
home, and some of them built *bothógaí* as shelters.

In the 1840s, despite regional differences in conditions, housing conditions
were at their worst, and the overall situation was accurately described by a
government commission as 'lamentable'. In parts of the west, almost half of
the houses had no chimney of any kind. Even in County Down, where
housing conditions were claimed to be the best in Ireland, almost a quarter of
the population lived in mud cabins with one room, while in County Kerry,
more than two-thirds of the people lived in these conditions.[9]

Contemporary discussions about the state of rural housing centred on the allocation of blame for such terrible conditions, and suggestions about what should be done. The answers given to these questions varied with the ideology of the observers, and often have striking modern parallels. They also show how deeply connected housing was with other aspects of rural life and attitudes. As in farming, a major division persisted between those who saw the ills of Irish life as reflecting deficiencies in the character of Irish people, and those who argued that the problems reflected the inefficiencies, or 'iniquities', of the economic and political systems in which country people had to operate.

One notion put forward was that the poverty was more apparent than real. People taking this position (who include some modern historians) argued that the Irish preferred a 'high consumption of leisure' to the laborious accu- mulation of goods, but that they did have enough wealth, for example, to operate an elaborate dowry system for their daughters. This is a common colonial stereotype. In Paul Scott's novel *A Division of the Spoils*, a memsahib living in India during the 1940s made her views clear.

> I sometimes think that the poverty is very exaggerated. Most of the Indians one knows could buy one up lock, stock and barrel … In the villages … every peasant woman has her gold bangles. No, no. It is not the poverty. It is the disease. The superstition. The *inertia*.[10]

As historians such as Joel Mokyr have pointed out, however, people who are dying of hunger are unlikely to be secretly hoarding wealth.[11]

The notion that the Irish were not really poor at all, produced a bracing 'Why don't they pull their socks up?' response. William Makepeace Thackeray dismissed the notion that Irish misery might be connected to the fact that many landlords were 'absentees', residing in England and living off rents collected in Ireland: 'Is the landlord's absence the reason why the house is filthy and Biddy lolls on the porch all day? … People need not be dirty … if they are ever so poor, pigs and men need not live together … The smoke might as well come out of the chimney as out of the door.'[12] Thackeray saw the landlord's role to be one of encouragement. He advised landlords to take their tenants 'good-humouredly to task' for any failings. Increasingly, however, the Irish people and nationalist leaders, blamed landlords for the appalling conditions endured by so many people. In 1839, a French traveller summed this view up. 'All the evils of Ireland, and all its difficulties arise from the same principal and permanent cause – a bad aristocracy.'[13]

18 Cottages in Adare, County Tipperary, photo courtesy Bord Fáilte.

As we have seen Irish landlords did not invest large amounts of money making their estates into the productive enterprises of the kind found in many parts of England. Their financial strategy was often to invest very little capital in developing their estates, but to live off the rents which their agents could collect from their tenants. Some small-scale projects by landlords to improve housing do deserve mention, because they show the conditions which were considered appropriate for the poorest tenants on estates. Estate farms sometimes included housing for workers, and estate workers and artisans sometimes benefited from attempts by landlords to prettify approaches to their demesnes. Striking examples of this can still be seen at Adare, County Tipperary (fig. 18).

The emphasis here was on picturesque exteriors. Internally, the houses were usually very simple, with a kitchen, parlour, and one or two bedrooms.

A very small number of landlords showed an interest in making more fundamental changes in working arrangements on their estates. The most dramatic instance of a landlord attempting to apply radical social ideas to restructuring an estate was the project undertaken by John Scott Vandaleur on the Ralahine estate in County Clare during the 1830s. Vandaleur was deeply influenced by the work of the British radical Robert Owen. He decided that his estate, a small one of just over 600 acres, should be re-organized as a 'commune'. In 1830, he had a row of stone cottages built for married couples, separate dormitories for single men and women and children, a large dining room and a meeting room. After some initial reluctance, tenants and workers on the estate supported the scheme. Vandaleur's agent, E.T. Craig, said that at first

> It was not easy to convince them of the advantage of united homes and combined social arrangements. The peasantry, though living in extreme wretchedness from their irregular employment and small earnings, were strongly attached to their old customs, and isolated, miserable cabins, with their apparent freedom.[14]

There was suspicion that the communal living quarters were really some kind of workhouse and in the first year of the experiment no-one came to visit the estate except a few tramps. After this, however, it became clear that the commune was doing well, and in the next year membership rose from 52 to 81 people. Six new cottages were built to accommodate the newcomers. Ralahine did not develop further, however. Vandaleur wasted his fortune gambling in London, and lost his estate in the late 1830s. The commune came to an abrupt end, bringing misery to its members, whose rights to land that they had previously held were not recognized by the new landlord.

Ralahine stands out because it was so unusual. Landlord involvement in housing schemes was more commonly confined to offering 'premiums' for home improvements to their labourers and tenants, either directly through an agent, or indirectly through grants given to the local farming society. In 1813, for example, the earl of Clancarty offered premiums for 'good windows, chimneys, floors, doors (hinges and locks included), painting, plastering and whitewashing'.[15] This kind of landlord activity did not have a major overall effect in the quality of housing on most estates, however. In general, landlords' involvement seems to have been indirect and distant. As in Irish rural society generally, the key social factor which influenced housing was probably the direct and often tense relationship which developed between labourers and farmers. For cottiers and landless farm workers, it was the farmer who employed them who controlled their living conditions.

A fundamental issue, which was repeatedly discussed in agricultural newspapers and manuals during the eighteenth and nineteenth centuries, was whether farmers should have any involvement at all in the provision of housing. In 1793, for example, an article in the *Irish Agricultural Magazine* attempted to convince farmers that they should encourage people to work as 'cottars' on their farms.[16] As we have seen in the previous chapter, the contract under which cottars were employed included the provision of housing as part payment for labour. The main advantage of having workers living on the farm was that labour would always be available to the farmer. Married workers with families could also be asked to provide extra help at harvest and other busy times. However, many people did not agree that having workers living on farms where they worked was a good thing. In the 1840s, for example, it was argued that the provision of cottages to farm labourers in Ireland had led to overpopulation of the countryside by people who could not hope to find permanent employment, but were exploited by farmers who rented cottages to them at inflated rents.[17]

The type of accommodation provided by farmers to workers varied greatly, ranging from hovels to 'substantial habitations'. Agricultural reference works of the early nineteenth century often included plans showing the best housing which could be provided. Most of these plans are of single storey buildings, built of stone, with slated roofs (fig. 19).[18]

However, the model houses illustrated in agricultural texts show the best that could be expected. For many farmers, the profit they could make from renting out cottages was their main goal. In at least some cases, as in County Cavan in the early nineteenth century, the farmers' main concern in planning cabins was that they should be located at outer parts of the farm, 'so the farmers may not be annoyed by them'.[19] The internal arrangements of model houses show a concern with 'decency' and to some extent with sanitation. Concerns with decency usually centred on sleeping arrangements. Many travellers in early nineteenth-century Ireland commented on the extent to which families had communal sleeping arrangements. In one-room houses, the entire family often slept on a single 'shake down' mattress spread on the floor. By the 1850s, however, this had become less common. Even very poor families were often able to provide separate beds or mattresses for adults, boys and girls.[20]

From the 1830s onwards, cottage plans generally had a small room or separate structure identified as a privy, but discussion of these was not detailed, and even less concern was shown with general sanitation and water supply. As late as 1907, Francis Joseph Bigger, one of the leaders of what has become known as the Celtic Revival, described sinks as 'unwholesome'. Washing up facilities in his plans for a model labourer's house were limited to a large

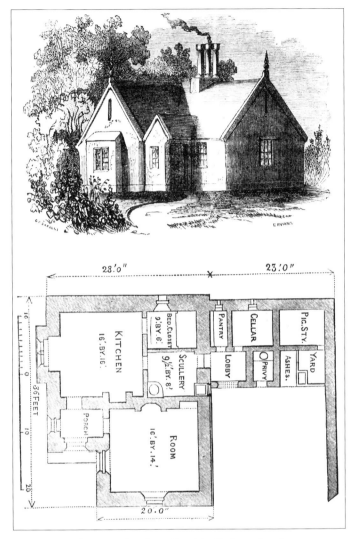

19 Cottage plan from Doyle's *Cyclopaedia*, 1844.

bucket or basin set on a stool or low table. After use the water was to be thrown out the cottage door (fig. 20).[21]

The subject which attracted most heated discussion, however, was not connected to the construction or plans of houses, but the question of whether land should be provided along with a house, as was usually the case when men were employed as cottiers. In the late eighteenth century, there was some discussion as to whether 'a race of little farmers' could be brought into existence, by encouraging self-improvement and investment in increasing

20 Interior of a cottage, by Francis Joseph Bigger, 1907.

acreages of land. Opinions were divided on the likelihood, or desirability of this.[22] In 1804 the *Annals of Agriculture* published an article warning against providing labourers with any permanent rights to land.

> One objection … is, the danger resulting from very small capital, that of an unwise penury … where land is allowed to keep one cow, if it begets a desire of rearing a second, both may be starved … A labourer' s fields should be chiefly confined to pasture, that the care of them may not interfere with his working for hire.[23]

The most commonly expressed worry was that labourers would put their main efforts into their own gardens, and proportionately less into their employers'. In 1825, however, the opposite was argued.

> There is a striking difference between cottagers who have a garden adjoining their habitations and those who have no garden. The former are generally sober, industrious and healthy, whilst the latter are too often drunken, lazy, vicious and frequently diseased.[24]

It was also argued that the produce of a garden would lessen pressure for increased wages.

> The produce of a garden … procures [the labourer] comfort and plenty, and by so doing, keeps within moderate bounds the wages of labour. Every man, who is averse to rising the wages of labour in husbandry, should at least encourage the culture of gardens'.[25]

The desirability of providing cottages with plots of land was also argued by an Irish Royal Commission in the 1830s, but as late as 1908, it was still being argued that supplying gardens, and even cottages, led to disrespect and a lack of diligence.

> If the labouring class were raised by means of education and grants for houses and plots of ground, it would be at the expense of the industrious farmer … The purpose of giving the labourer an acre of land was that it should supplement his work as a labourer … But having it he could afford to do with less labour than before.[26]

The question of providing land for labourers revealed deep tensions between tenant farmers and their workers. Farmers aimed to keep control of rents paid for land by labourers, and some were even prepared to use the Land League, the body which was successfully campaigning to transform tenant farmers into owner-occupiers, to achieve this. In 1880, for example, a landowner in the Shanagarry area of County Cork began to let small patches of ground to some poor labourer-fishermen at rents below those that they had been paying to local farmers. In response, the farmers tried to intimidate the labourers into giving up their plots. One tactic they adopted was to set up a branch of the Land League which they hoped could be used to pressurise the labourers more effectively. At the inaugural meeting of the branch, however, the labourers took over the meeting in protest. Charles Stuart Parnell, the charismatic leader of the League intervened, and attempted to extend the work of the movement to include the interests of farm labourers.[27] The Land League began to build houses for homeless, especially evicted people.

The houses were very basic, but most were better than the cabins that the people had lived in before, and they were certainly better than the workhouse. This work, however, was overtaken by legislation. In 1883, Parnell and the Irish

21 Cottage built by the Congested Districts Board, 1905
(Welch Collection, Ulster Museum).

National League managed to have an act passed at Westminster under which local authorities were empowered to build labourers' cottages. The act was modified in 1892, and houses built under this and later acts led to the construction of small, decent houses, which are still an important feature of the Irish landscape (fig. 21).

Paradoxically, however, as state action to provide housing became more effective, agricultural labourers were rapidly decreasing in numbers. Mass emigration continued in the decades following the Famine. Between 1850 and 1900, more than four million people left Ireland, mostly the rural poor. The number of people living in one roomed houses decreased proportionately, from 43.6% in 1841, to 1.9% in 1901.[28] For those who remained, things began to improve, with the help of government grants given out by bodies such as the Congested Districts Board, and the Irish Land Commission. By 1900, provision of houses by rural district councils was also having a major impact. In 1921, around 54,000 agricultural labourers' houses were either planned or being built. Most of these were in Leinster and Munster, where farms were largest and labourers still relatively numerous.[29] Building has continued ever since. During the 1920s and 1930s, roof tiles were sometimes used instead of slate, cement plaster instead of lime plaster, and steel replaced the earlier wooden window frames. From the 1950s, entrance halls became common, internal sewage systems were installed, and internal water supplies began to supersede communal outdoor pumps. During the 'quiet revolution' of the 1950s, many houses in the countryside were also supplied with electricity. By 1961, 400,000 homes in rural areas in the Republic and 23,000 in Northern Ireland had been provided with electricity.[30]

Terrible rural poverty still existed in many parts of rural Ireland well within living memory, however. In one survey carried out in Northern Ireland in 1947, living conditions which most modern urban dwellers would find shocking, were quite common.

The most spectacular instance of poor housing was recorded in north County Antrim.

> One house differed from all the others: this was simply the four walls of the last house in a former clachan. There was no roof, no windows, no door, but one person occupied it as a shelter, sleeping with his head in a round bottomed three legged pot and feet in a barrel. Any meals cooked must have been made in the pot during the day, but food was mostly obtained in a neighbour's house.[31]

This was an extreme example, but many services which are now taken for granted were not at all widely available even after the Second World War. The situation in County Fermanagh with regard to water supply, for instance, would have been typical of many rural areas.

> Despite the abundant rainfall, the numerous lakes and many streams, the greatest difficulty in Fermanagh is the provision of water. It is truly a case of 'water, water everywhere nor any drop to drink'. Piped water is almost unheard of for the rural dweller whether farmer of labourer. Drinking water is everywhere obtained from wells, which are mostly shallow and dry up in summer. Many farm houses in the limestone country are dependent on rain water supplies … Typical of the water supply position was the empty farm house … occupied by the survey party in June 1944 … Rough [lake] water for washing had to be fetched from under a bridge a quarter of a mile away, while water for drinking and cooking was carried in buckets a full mile from the nearest well.[32]

Sanitation was no better. In a survey of over 7,400 houses in Fermanagh, it was found that only 4% had mains sewage, 27% had dry toilets, and 69% had no provision of any kind.[33] Fortunately, since the 1950s, things have improved dramatically. Major housing schemes in rural Ireland, north and south, have ensured that there is now little difference between public housing in town and countryside.

FARM BUILDINGS

Big Houses and their demesnes became dominant features of the Irish landscape during the eighteenth century, and the earliest surviving planned farmyards were associated with these. As in housing, however, contrasts were very great. Until the mid-twentieth century, it was still easy to find conditions which would seem very primitive to modern Irish farmers. Travellers and agricultural writers during the eighteenth and nineteenth centuries often commented on the lack of separation between animals and people which they

22 Byre dwelling from *Machaire Gathlán*, County Donegal, rebuilt in the Ulster Folk and Transport Museum: (a) exterior; (b) bed, dresser and grate; (c) cattle stalls.

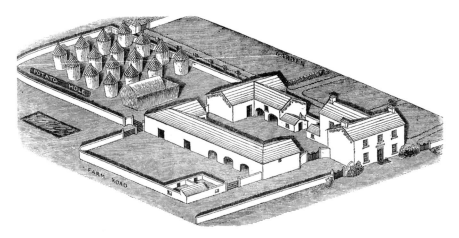

23 Farmyard on the estate of the earl of Westmeath, *Irish Farmer's Gazette*, 1846.

found in Ireland. As in western Scotland, many small farmhouses in the west of Ireland were 'byre dwellings'; people lived at one end of a room, with cattle tied at the other (fig. 22a, b, c).

This arrangement was not always due to poverty. In northern Europe, for example in Denmark, quite grand houses were constructed with a living space at one end and cattle stalls at the other. However, in Ireland the arrangement seems to have been found mostly on small, very poor farms. In Ireland, as well as Scotland, byre fittings included a 'sink' or drain to carry away liquid waste, and the area on which the cattle were tethered had a stone floor, except the place where the animals stood. This had a clay surface, to make it more comfortable for animals to kneel down.

One practice which became part of the stereotype of the stage-hall Irishman, was the close living relationships between some families and the farm's pig. This arrangement was commonly referred to as the Pig in the Parlour, and was frequently used in caricatures. In a situation of very limited resources, however, it made sense. Pigs flourish in the same temperatures as people, and especially when a sow was producing a litter, moving her into a warm space meant that she could give birth in comfort and safety. As with byre dwellings, however, where practicable, pigs were kept separately. Arthur Young mockingly recorded efforts by the rural poor to create separate spaces for their pigs in the late eighteenth century.

The luxury of sties is coming in Ireland, which excludes the poor pigs from the warmth of the bodies of their master and mistress. I remarked little hovels of earth thrown up near the cabbins, and in some places they build their turf stacks hollow, in order to afford shelter to the hogs.[34]

From the late eighteenth century, Irish farming texts began to include plans for model farmyards (fig. 23). These typically included stables, byres, pig houses, and barns. By the mid-nineteenth century discussions about lay-out and internal arrangements were very detailed, often taking both efficiency of operation, and the comfort of livestock into account. Measurements and building materials were given in detail. For example, in 1839, the Irish agriculturalist, John Sproule, recommended changes to the internal construction of cattle byres.

> The common method of attaching cattle in these houses is to upright posts placed in a row, at the distance of about two feet from the wall. Round each post is a movable ring, to which is attached a chain which passes round the neck of the animal. When attached in this manner, the usual distance between posts is about four feet. The animals feed from a low manger, formed for the most part of a raised edge of stone, between which and the wall, the food is placed.[35]

Sproule recommended that it would be better to give each animal its own stall, separated by partitions of stone or board.

Particular care was advised in the stabling of horses. John Sproule explained that 'The horse is an animal in a highly artificial state [produced by selective breeding], and requires to be treated with a degree of care beyond that bestowed on any other quadruped'.[36] A contemporary, Joseph Lambert, described how this should be reflected in the construction of stables.

> [Stables should be well-lit, as horses] will not only get shy, but blind from sudden exposure from a dark stable … Lofty stables will contribute much to the health of horses, and a stable should not be less than twelve feet high … It is an advantage in many respects not to have a loft, as grooms are then less inclined to have hay constantly before their horses; and it is for this reason that trap-holes over the rack should never be allowed for the purpose of letting down hay, but make the groom fork it in at certain times … and not leave it to be blown upon. Hay is also

better and fresher from the rick than when lying on a loft to must: such holes also cause dust and dirt to be blown about the heads and ears of the horses, as well as cold wind coming in when the loft is empty.[37]

Another outbuilding where the need for even more sensitive adjustment and control of conditions was recognized, was the farm dairy. John Sproule recommended that the dairy should have 'Equality of temperature during every season of the year, and frequent renewal of the air, so as to have it fresh and sweet … The situation of the dairy should be shaded by trees, to prevent the great increase of temperature which is produced by the rays of the sun.' On smaller farms, a milk-room at least was recommended for storing milk before churning.

> 'It should have its windows to the north, and be so formed as to preserve a cool and equal temperature. The windows should be formed of gauze cloth, which may exclude flies, but admit the air, and protected from mice and accidents by iron bars and a grating of wire. This apartment should be kept cool in summer, but in winter heated by a stove or otherwise, so as to maintain a temperature of 50° to 55°.[38]

It was commonly alleged that in Ireland, even gentry who had resources to build model yards, preferred more conspicuous spending on their houses, which were too often 'built for show more than utility … frequently the cause of embarrassment.'[39] But despite these ongoing complaints, planning of farm yards did develop. By the mid-nineteenth century, Acts of Parliament enabled farmers to apply for grants for constructing all major buildings such as barns, cow sheds and stables, as well as root houses, boiler houses, manure sheds, and poultry houses. By 1900, one very important advance was in the provision of hay sheds. By this time hay was the major crop in Ireland, and given the country's damp climate, farmers had a strong incentive to store the dried crop under cover. The construction of open-fronted hay sheds, with corrugated iron roofs and sides, was grant-aided, and by the 1900s these 'Government barns' were claimed to be more common in Ireland than in Britain (fig. 24).[40]

Development did not always mean that farm buildings got bigger. When flails were used for threshing, well-planned barns were made big enough to allow space for flailing grain, and also to store a stack of unthreshed grain, which might be threshed over a number of days, or even weeks. By the 1830s,

24 A 'Government' barn, built at Sentry Hill, County Antrim (Dundee collection).

however, the introduction of mechanical threshing machines speeded up the operation of threshing. As a result, it became common for the crop to be stored in outdoor stacks, and threshed a stack at a time. If a barn threshing machine was installed, the grain about to be threshed could be moved to a compartment above the threshing machine, and fed down as needed. Portable steam, and later tractor-driven, threshing machines almost always operated in the open air. When these became common from the later nineteenth century on, a farm's entire grain crop could often be threshed in a single day, or several days at most, so unthreshed grain did not have to be moved into a barn at all.

Barns were given a central position in many early plans of farm yard lay-out. A lot of livestock fodder, and straw for bedding, were prepared in the barn, and it made sense for it to be as near as possible to buildings where animals requiring these were housed. It was also pointed out, that since a major function of yards was to provide shelter for cattle, and since the barn was usually the highest building in the yard, it should be placed on the side of the yard from which prevailing winds blew. Hay sheds, on the other hand,

25 Farmyard from Coshkib, County Antrim, rebuilt at the Ulster Folk
and Transport Museum.

were thought to be best positioned so that the open front faced south, for
maximum heat.

Plans were published for farm yards which were circular or oblong in plan,
but the most common lay-out was around a square or rectangle, or on smaller
farms, in a line extended on either side of the dwelling house (fig. 25).
To a large extent, the arrangement of these buildings was influenced by
available space and the contingencies of the site. At their best, their lay-out
shows the same sensitive adjustment to the small-scale local environment,
which also characterized Irish houses, farming techniques, and general ways of
life, before these were obliterated, for good and ill, by the forces of modernization, standardization and globalization.

CHAPTER THREE

Ridges and Drains

ETWEEN THE MID-EIGHTEENTH century and the mid-nineteenth
century, a lot of hilly land and bog land was brought into cultivation in
Ireland. These projects are important for this book, because they were sometimes recorded in detail, and show contemporary notions of what constituted
viable farms, and the techniques that were available to create them. The work
of establishing such farms involved some of the richest landlords and their
poorest tenants.

The Devon Commission's report, published in 1847, recognized three main
approaches taken to land reclamation by Irish landlords:

1   The landlord could organise the entire reclamation scheme, and only
    after it was complete, let the land to tenants.
2   Tenants could be allowed to settle on marginal land, to reclaim it,
    with some assistance from the landlord. This assistance might include
    cash grants or loans, training in approved techniques, or could be
    confined to the inducement of several years' tenancy at a lower rent.
3   The landlord could leave the whole operation to the unassisted labour
    of his tenants. Most contemporaries, and modern historians, agree
    that this last approach was the one most commonly taken by
    landlords, until the mid-nineteenth century.[1]

James McParlan, assessing the possibilities for land reclamation in County
Donegal in the early nineteenth century, found the principal obstacles to be 'a
want of capital and knowledge in the poor, and of spirit and enterprise in the
rich'. The dependence of any scheme on the labour of the rural poor was
recognised by McParlan when he commented, 'Emigration is a great obstacle
to reclaiming bogs and mountains. In the course of last year, upwards of 4,000
persons left the single port of Derry.'[2] By the 1820s, however, the piecemeal
reclamation of the west of the county was well under way: 'A great deal of the
Boylagh is reclaimed and cultivated, there being little other soil … New cabins
are rising in many places, and would still more rapidly if roads were open.'[3]
The relationship between road-building, drainage and land reclamation was

widely recognized. George Cecil Wray, a Donegal farmer who testified to the
Devon Commission, was emphatic:

> There are immense tracts of improvable moorland totally neglected on
> two of the largest estates in this district. Many thousand acres, now only
> yielding a very small return from pasture, could, without any further cost
> to the proprietor than making roads through it, be made valuable land.[4]

One of the most publicized reclamation schemes in Donegal was carried out
on the Cloghan estate in Glenfin, under the direction of the estate's agent,
John Pitt Kennedy. The estate was 16,000 acres in area, and in the 1820s about
14,000 of these were described as waste. Kennedy took possession of all the
farms on the estate, which had been divided in rundale strips by the tenants.
Kennedy abolished rundale, and laid out new compact holdings. About 160
new farms were situated on mountain land which was considered capable of
improvement.[5] It was claimed that Kennedy tried to give erstwhile tenants the
same amount of arable land as they had previously held in rundale, and
compensated those moved on to unreclaimed mountain. In cases where the
arable land was considered insufficient to support a family, 'the claimant was
placed on a wasteland farm, of improvable land, and dimensions suited to his
capability – averaging twenty acres; and he besides received some compensation from those among whom his former small arable lots may have been
divided.'[6] Tenants of mountain farms were allowed to live rent free for between
three and seven years, after which time the rent was gradually increased.
Assistance was given with building new field walls and cottages and by 1841
twenty miles of road had been built through the estate. Kennedy also set up a
winter loan fund. Money from the fund was to be used for 'some reproductive
object', such as liming land, or the purchase of new farm implements or a cow.
To qualify for a loan tenants had to carry out land drainage on their holdings,
under the supervision of an 'agricultural teacher'.[7] The drainage project undertaken on the estate was described at length by the Devon Commission. One
interesting aspect of the work was that it combined the use of thorough drains,
an 'improved' technique, with the ancient technique of making cultivation
ridges. The ridge and furrow pattern provided surface drainage in the early
stages of reclamation.[8]

The teacher who inspected the drainage work on the Cloghan estate was
employed on an agricultural school opened there in 1837. The school on the

Cloghan estate was smaller than the well-known Templemoyle School in Derry, but attempted the same synthesis of practical and academic education. The surrounding moorland and bog was seen an advantage of the site.

> The object in view of the establishment of [the school] … was to qualify young men to fill the situations of capable land stewards, agriculturalists, and agents of estates; also teachers of agricultural schools. And the mode adopted was to select … [an estate] which, consisted almost entirely of waste land, when active operations were about to commence with a view to their reclamation and improvement in all requisite details.[9]

Girls were also taught at Cloghan, the object being 'to render them good servants'. The well-known English travellers, Mr and Mrs S.C. Hall, who visited the estate, were impressed by the commitment of Captain Kennedy and his sister:

> It is surely enough to say of the boys' school, that while Miss Kennedy receives every day from the mistress a report of each pupil's progress, and inspects the school herself several times during the week, her brother … watches over the boys with the deepest solicitude.[10]

CULTIVATION RIDGES

Many of the differences in techniques used in reclamation can be linked to the resources available to particular farmers. All land 'reclamation' required intensive labour, however, and some techniques were used on large and small schemes alike. Reclamation required digging or ploughing, liming, and in some cases paring and burning. One method used by the smallest farmers, but also found in many large-scale projects, was the use of high, narrow cultivation ridges.

Cultivation ridges were widely used in breaking in marginal land for crops, and played a central role in Irish tillage systems until well into the nineteenth century. Ridges could be made using either spades or ploughs, and we will outline some techniques of ridge making in the chapters dealing with those implements. Here, however, we will look at their usefulness in reclamation of land.

Ridge-making techniques were cited by many early agriculturalists as evidence that Irish agriculture was backward and inefficient. At Forth, in County Wexford, for example, Arthur Young found that one third of each

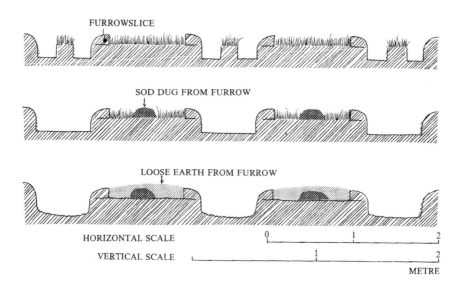

FURROWSLICE

SOD DUG FROM FURROW

LOOSE EARTH FROM FURROW

HORIZONTAL SCALE

VERTICAL SCALE

METRE

26 One method of constructing a lazybed.

ridge was left unturned during initial ploughing. He condemned this as 'execrable'.[11] Even Young, however, recorded that some 'improving' landlords were using these 'lazy-bed' techniques and finding them satisfactory. The practice of building up ridges on strips of untilled land was regarded as being particularly associated with Ireland. The widths of these strips varied. Ridges built up on strips which had been deliberately left unturned were known by various terms, including *iomairí*, rigs, or lazy-beds (fig. 26). At Killarney, for example, Arthur Young noted that, 'Mr Herbert has cultivated potatoes in the common lazy-bed method, upon an extensive scale, and he is convinced, from repeated experience, that there is no way in the world of managing the root that equals it.'[12] By the early nineteenth century, some observers were claiming that larger crops of potatoes could be produced from lazy-beds than from the newer system of drill cultivation. The growing acceptance of lazy-bed techniques arose chiefly from the recognition of their usefulness in breaking in waste land for cultivation, where one potato crop could bring 'certain descriptions of ground into profit'.[13] Lazy-beds were particularly suited to the small-scale, labour-intensive methods available to cottars and small farmers, when breaking in marginal land. They were also used in some large-scale reclamation schemes, however. In 1846, W.S. Trench planted 100 Irish acres of potatoes on mountain land in Queen's County (Offaly). The land was limed, and the outer edges of the lazybeds turned by ploughing. The beds were then manured using guano, and covered with clay dug out of the furrows. Trench found that, 'The Potato grew to perfection in this rude description of tillage; and whilst it was growing, the heather rotted under the influence of the lime, and, together with the superabundant vegetable matter, was turned by the action of the lime into a most valuable manure'.[14] The advent of potato blight in 1845, however, brought a halt to this technique of reclaiming waste. This was lamented by the Devon Commission: 'The Irish or lazy-bed method of planting potatoes, supplied the most minute conceivable system of artificial draining for that one crop'.[15] The sensitive adjustments practised within these techniques in response to aspect, soil type, crops grown, their place in the crop rotation, and the methods of harvesting, show that, although developed by the poorest farmers, ridge-building was part of a remarkably sophisticated cultivation system.

The acceptance of lazy-bed techniques as efficient was not so widespread in long-established tillage areas. By the beginning of the nineteenth century, some Irish farmers were experimenting with techniques of cultivation which did not involve the construction of ridges of any kind. One criticism of ridge cultivation was that the land taken up by furrows went to waste. As noted earlier, furrows could cover between one half and one third of the cultivated area. It was also claimed that on rounded ridges the plants growing on the sides facing away from the light were 'shaded', and this hindered growth.[16] Tighe reported in 1802 that some farmers in County Kilkenny had attempted to cultivate wheat on level ground. This had not, however, been a success:

> This method was not found to answer as well as the common one: by means of the ridges in which wheat is usually placed, a greater surface of corn is exposed to the influence of the atmosphere, the ears are not so sheltered by each other from the air, nor so shaded from the light as they would be if the field was even: that the air should circulate freely about the ears, is known to be an essential point with regard to wheat; a contrary situation encourages blights and mildew; … and though corn might not grow in the furrows between the ridges, yet the roots receive nourishment from them.[17]

These arguments for and against ridges, however, dealt with aspects of the system which were to some extent secondary. The central aims in making

ridges were to achieve drainage, and to deepen soil. The first major impulse to
level ridges, therefore, came with the spread of the large-scale schemes of
thorough drainage. By the 1840s, one of the leading Scottish experts on
drainage, James Smith of Deanston, believed that his drainage system rendered
ridges obsolete:

> [Smith] alluded to the practice, which he said had existed from time
> immemorial, of throwing the land into ridges and furrows, and showed
> that, by the soil being washed from the tops of the ridges into the
> furrows, the higher parts of the field produced comparatively little crop
> … In thorough-drained land no drop of water should run on the surface
> in any direction, but should penetrate into the ground where it fell …
> Wherever the land was drained, it was necessary that the high ridges
> should be done away with, and the land laid perfectly level.[18]

In Ireland, however, the application of Smith's methods was not always
accompanied by an immediate levelling of cultivation ridges. Instead, ridge
and furrow systems were integrated with thorough-drainage to increase the
latter's efficiency, especially in wet, marginal land. On the Glenfin estate in
County Donegal, for example, the laying down of drains was followed by the
building up of ridges five feet wide, separated by furrows of two feet. If the
land was flat, the ridges were made to run at right angles to the drains. If the
drains ran down a slope, the ridges were made at an oblique angle.[19]

Some cultivation ridges, especially those made by spades for planting
potatoes, were deliberately given firm, steep sides. These sides, known as
bruacha, brews, or brows, were recognized by some observers as possibly
hindering the efficient operation of thorough drains: 'The loosened soil at the
bottom of every potato trench becomes necessarily the line of filtration, and
the harder brows forming the edges impede the lateral transit of the water to
the drains, so that, in fact, a good deal of the filtration in that case would
resemble what takes place in an undrained field.'[20] The solution suggested was
not that the drains should be flattened, however, but that, as in the Glenfin
example just given, they should be made to run at right angles to the drains,
and that the bottom of each furrow should be made to slope downwards, so
assisting the flow of water.

One factor leading to the short-term retention of cultivation ridges on
recently drained ground was their use in achieving subsoiling. The breaking up

of impervious clays below the land surface was a central part of Smith of Deanston's system. Smith devised a subsoil plough to achieve this, but Irish contemporaries claimed that the implement was not always effective. James Kelly, agriculturalist to the Longford Farming Society, was emphatic: 'There is no tenant that I know of in this country that would think of getting a subsoil plough, nor do I think it would be wise of them to do so.'[21] In 1845, Lambert summarized the reasons for this lack of enthusiasm:

> The subsoil plough, as now in use, was invented by the justly celebrated Mr. Smith, of Deanston ... but, though admirably adapted to good flat plough-lands, such as the carses of Scotland, [the implement] is suitable here only in similar cases, and to proprietors and extensive farmers who keep strong horses and powerful teams. It is altogether out of the reach of our peasantry; and there are few large farmers who are so systematically extensive in tillage as to have the horse power necessary for the work.[22]

Agriculturalists recognized that subsoiling could be achieved by spades and spade ridges, which were more suitable to Irish conditions. An impervious substratum of clay, commonly known as 'lacklea' in many areas of Ireland, could be exposed and broken up in the process of making ridges. Ridges, therefore, could have the same effect as a subsoil plough, and produce equally striking results. In County Sligo in 1823, for example, it was found that, 'Some of the low and moory grounds with ... (lacklea] within three or four inches of the surface, are rendered fertile, and free from stagnant water by being cultivated for a potato crop, the subjacent stratum being carefully cut through by the loy or spade'.[23] The practice of 'reversing' ridges from year to year was well established by the mid-eighteenth century. This meant that when ridges were made on a piece of land for a second year, the new ridges were built over the previous year's furrows. Agriculturalists pointed out that by this method, subsoiling could be effectively achieved over a whole field within two or three years.[24]

Despite all the advantages claimed for ridge cultivation, however, the building up of high, narrow ridges on a large scale did decline in the later nineteenth century. There were several important factors leading to this. The integration of ridge-making and thorough-drainage was most effective when marginal land was being broken in for cultivation. Once land was fully reclaimed, however, the need for ridges diminished. Other changes in agricultural practice also led to the slow disappearance of ridges. These included the

27 Planting potatoes on ridges near Derrylinn, County
Fermanagh.

spread of drill machinery in potato cultivation, and the adoption of scythes
and reaping machines as grain harvesting implements. The implications of these
changes will be discussed in later chapters. For the Devon Commission, reporting
in 1847, the main disadvantage of ridge cultivation was that the intensive labour
required would have been better employed in constructing thorough drainage
systems, so creating much more permanent improvement of land:

> We may certainly estimate the extra labour on an acre of lazy-bed
> potatoes, as compared with drill culture, at £1 10s., and this extra labour
> on two acres would have been nearly sufficient, on the average, to have
> thorough-drained one acre; assuming, therefore, the quantity of ridge
> potatoes in Ireland to have been one million of acres, the extra labour
> wasted in this crop annually would have permanently drained nearly half
> a million of acres per annum, or the labour thus wasted during the last
> twenty years might probably have drained all the productive land
> requiring drainage in Ireland.[25]

During the second half of the nineteenth century, all of the factors which
made ridge cultivation less attractive to farmers grew in importance. However,
even in the new millennium, some small holders cultivating tiny potato patches
still rely on ridges to produce excellent crops on shallow, poor soils (fig. 27).

DRAINAGE

By the late eighteenth century, some very large-scale schemes to improve land were undertaken by landlords. Arthur Young, for example, gave the title 'the Great Improver' to Lord Chief Baron Forster, of Cullen, after inspecting a scheme which brought 5,000 acres into culivation. This land, which had previously been 'a waste sheep walk, covered chiefly with heath and some dwarf furze and fern', had been ransformed into 'a sheet of corn'. The improvement had required the erection of twenty-seven lime kilns, and the construction of hollow, underground drains required 'the burying in ... of several millions of loads of stones'.[26]

Arthur Young recorded several examples of rivers being widened, or navigable channels cut through bogs. One spectacular scheme was undertaken by Robert French of Moniva Castle, County Galway. Before drainage, the castle was surrounded by 'two very high, red, deep wet bogs, impassable for any beast of burden, very difficult for even men to pass'. The bog on the east side of the castle had grown up so high that it blocked the view from ground-floor windows. A river between this bog and the castle overflowed during heavy rains. There were some small drains already cut through the bog, but in 1774 French had them widened and deepened:

> The sides of the drain were so high, that I was obliged to [have them] cut ... in some parts into benches, in the form of stairs, to prevent the men at the bottom from being overwhelmed, which would once have happened, only that a man standing on the surface, observing the bog to burst, gave the alarm, by which he saved the lives of several men; for in a few moments many perches in length of the drain were filled up to the top.

French had the river flowing near the castle diverted, and so produced ten acres of meadow. The new river course had a fall which was used to power a bleaching mill, and another drain cut from a nearby lake was used to power the mill in summer. Some of the wooden piles used to hold the new river banks in place were washed away during heavy rain, but these were replaced by stones carried by boat. Boats were also used on the broadest drains cut through the bog. One drain was cut so that boats could carry manure right in to the castle's farm yard. The boats also carried limestone and manure around the bog, to points from which they were spread as fertilizer. French claimed

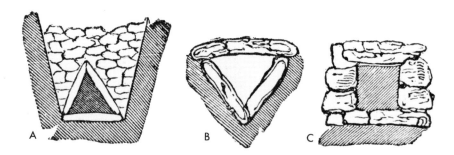

28 Three methods of constructing stone drains, from Martin Doyle, *A cyclopaedia of practical husbandry* (London, 1844), p. 204.

that when the scheme was completed, not only was the land greatly improved, but the bog had subsided fifteen to twenty feet, and no longer blocked the view from the castle's windows.[27]

Young recorded the construction of underground 'Hollow' or 'French' drains (fig. 28), not only under the direction of great landowners, but by some smaller farmers as well. At Kilrue, for example, he found that 'Hollow, called French drains are very general, even among the common farmers: some done with stones, but much with sods, laid on edge in the ground, they dig them 2½ or 3 feet deep'.[28] His note on drainage in the Barony of Forth in County Wexford, is probably more typical, however: 'They drain only with open cuts, no hollow ones done'.[29]

By the beginning of the nineteenth century, agriculturalists were advocating scientific 'thorough' covered drainage as essential for good husbandry. The methods developed by the Englishman, Joseph Elkington of Warwickshire, were generally agreed to be the best available. A key part of Elkington's technique was to release underground springs, using an auger or borer (fig. 29). A typical application of the method was recorded in County Dublin, in 1801:

> George Grierson, Esq. has had some ground drained in the Parish of Tallaght, this year, according to Mr Elkington's mode, which appears to answer perfectly well; common drains were first opened to a depth of about two feet, and then bored at every eight or ten yards with a three inch auger; about six or seven feet deep; some of the holes, when I saw them, were throwing off a quantity of water, while others were dry, but I was informed, had run.[30]

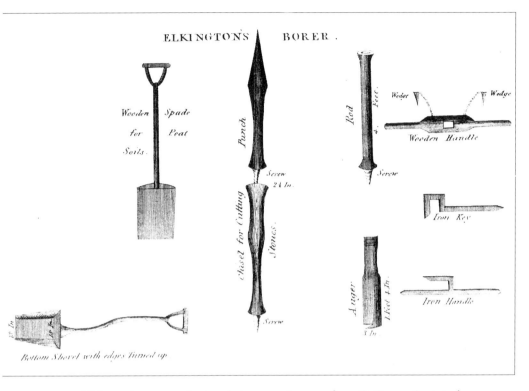

29 Elkington's auger, and other drainage equipment, from C. Coote, *Statistical survey of the county of Cavan* (Dublin, 1802), p. 118.

Another account, published in 1802, claimed that Elkington's methods had been 'tried in several parts of Ireland, and always with great success'.[31] Despite these projects, however, observers in most parts of Ireland lamented the lack of application of new principles. McEvoy's comments on County Tyrone in 1802 probably described common practice throughout much of the country: 'Open drains are in common use, only temporary to save crops in most situations, when the latter end of the spring is wet. The secret of hollow draining is very little understood in any part of the county, much less the intercepting, or cutting off springs. Sod drains are not known'.[32] Hely Dutton, writing over twenty years later, repeated criticisms of what he saw as the haphazard drainage methods employed by most Irish farmers. He also warned, however, that where systematic drainage was attempted, proprietors should ensure that they engaged a bona fide expert:

A few years since, Mr Elkington, nephew to the celebrated drainer in England came over to this country under an engagement to a few spirited gentlemen ...; this gave such an impulse to draining that hopes were very generally entertained that great and permanent advantages would accrue to Ireland; certainly his works were excellent, and more neatly executed than the general practice had been, but [his methods] ... were little or nothing different from those in practice by every other scientific drainer ... Since that period many itinerant quacks have started up in this branch ... whose low terms have blinded the judgement of some landed proprietors, that in this, as in landscape gardening, have mistaken neatness of execution for correctness of design ... Mr Hill, a native of north Britain, followed Mr Elkington; he was imported by the Farming Society of Ireland, and as far as I can judge ... is an excellent drainer ... since that period I have not heard of any person of eminence.[33]

By the 1840s, the drainage system advocated by James Smith of Deanston[34] was becoming widely known in Ireland. The Devon Comission found that on most large-scale schemes, Smith's plan of closed drains placed twenty-one feet apart, at a depth of two feet six inches, was being followed. Small drains, running parallel to one another, fed into larger 'sub-drains', which in turn discharged into main drains. Smith's system was not always followed rigorously, however. One deviation commented on by the Commission was the practice of running drains diagonally across slopes, rather than straight up and down them. The Commissioners were confident that this would soon be recognised as 'erroneous', as 'it admits of a simple mathematical proof.' The diagonal layout was defended by some writers, however, who argued that on steep inclines, drains running straight down the slope would produce a current so strong that it would carry fine top soil away, and also possibly overdrain the ground.[35]

Smith's parallel lay-out of drains at a fixed depth was intended to be applied without alteration to any land, irrespective of small local changes in slope, soil type, or moisture content. The modified version of the system, devised by Josiah Parkes, which entailed fewer drains, laid deeper, also used a rigid parallel lay-out. It was this rigidity, rather than a rejection of the recommended techniques of drain construction, which led to a decline in the use of both systems. The point was clearly made in an Irish agricultural text published in

1863. 'People began to see that, owing to the diversified character of the soils, subsoils, and substrata of the land in this country, it was impossible to fix any one uniform depth or distance apart for the drainage of soils of every description.'[36]

By the late nineteenth century, attempts to produce a completely standardized system had given way to the notion of 'occasional drainage', which took account of small variations in soil, slope and waterlogging within the area to be drained. This represented, in a more systematic form, something of a return to the approach associated with Joseph Elkington. It had been claimed that Elkington had become a famous drainer, not so much because of his scientific mastery of the principles of drainage, but because he was intuitively aware of the individual requirements of any area on which he worked. Similar sensitivity began to be expected of more modern drainers.[37]

One important change in drainage techniques, which occurred in the mid-nineteenth century, was the increasing use of clay tiles and pipes in the construction of drains. The earliest tiles to be widely adopted were known as 'horse shoes', which were placed on 'sole' tiles laid along the bottom of the drain.[38] In 1843, however, John Reade exhibited machinery for making cylindrical tiles at Derby Agricultural Fair.[39] These tiles quickly came into widespread use, and by 1870 were regarded as the most common type in Ireland:

> Pipe tiles are the materials now most extensively used. They have been made of various sections, such as circular, horse-shoe shape, and oval. The section of pipe now most universally used is the circle. The diameter of the bore varies from one inch upwards. In the east of Ireland and where rainfall does not exceed thirty inches per annum, pipes an inch and a half in bore are quite capable of carrying off the water of parallel drains. In the southwest of Ireland, where the rainfall is double this, or where it falls in very heavy showers, two inch pipes should be used.[40]

By the end of the nineteenth century, the dominance of cylindrical tiles seems to have become complete. A British text, published in 1909, declared that 'horse-shoe' drains were 'a thing of the past'.[41]

After the Great Famine of the 1840s, large-scale drainage schemes were backed by government legislation and monetary grants. A Drainage Act had been passed in 1801 but proved ineffective. The Act of 1847, however, was much more successful. By 1850, over £3,500,000 in loans had been applied for under the terms of the Act, 20,000 labourers were employed in drainage, and 74,000 acres had been reclaimed.[42] The trend towards large-scale organization increased during the later nineteenth century. The Famine and its aftermath had led to the removal, by starvation, emigration and eviction, of huge numbers of the poorest farmers, who had been the main agents of land improvement during the early part of the century.

CHAPTER FOUR

Manures and Fertilizers

PARING AND BURNING

PROCESSES THAT ENRICH the fertility of soil could take place either before or after tillage. One practice frequently discussed in eighteenth-and nineteenth-century agricultural texts was paring and burning. There were many variations in the techniques used in this process, but the basic operations were stripping off the surface sod, allowing it to dry, burning it, and then spreading the ash on the soil as a fertilizer. In the early eighteenth century the practice was widely condemned by landlords, and in 1743 the Irish Parliament passed an act forbidding it.[1] The Act does not appear to have been very effective, however, and by the early nineteenth century condemnations had become a lot less sweeping. The change in attitude came with the recognition that, as with lazy-bed techniques, controlled paring and burning could considerably speed up the reclamation of marginal land. Arthur Young, though often critical of the practice, made use of it in the reclamation scheme he organized on the southern side of the Galtee mountains.[2] Burning some types of heath, along with the top sods containing their roots, could reduce previously tough, poor vegetation to productive ash. By 1802, Charles Coote, like some other observers, concluded that, 'The effects of burning land were not well understood when the legislature imposed the heavy penalty against this process'.[3] It was argued by some agriculturalists that it was not paring and burning which ruined the land, but the over-cropping which often followed it. In 1807, Rawson, describing the practice in County Kildare, made this point clearly:

> The common practice of burning the whole surface [of upland] and then applying the entire ashes on the remnant of the soil, taking three or four exhausting crops, cannot be too much reprobated; by it the land is completely exhausted, and men say how injurious paring and burning is, not considering, that the injury lies in making an improvident use of ashes.[4]

In 1810, Horatio Townsend argued that the unattractive appearance of the land after paring and burning was often the source of mistaken claims that the process was injurious:

> Were [paring and burning] ... attended by the evil consequences so
> frequently deplored, the lands of Kinalea [County Cork] would by this
> time be reduced to a state of infertility. The contrary, however, is the fact
> ... It is very probable that the naked appearance of land, let out without
> grass seeds after burning, has been a principal cause of objection to the
> mode. But this barrenness is more apparent than real.[5]

The county statistical surveys published in the early years of the nineteenth
century contain several descriptions of the use of paring and burning by
landlords, some of whom had ploughs specially modifed for the purpose.[6]
Modified forms of the process were also used in some of the most systematic
schemes of land reclamation. In 1835, for example, one model project at
Tullychar, County Tyrone, included the paring of sods, using the local type of
breast-ploughs, known as flachters (fig. 30).[7] Paring and burning continued
throughout the later nineteenth century, but appears to have become increasingly
confined to small farms in western counties such as Donegal[8] and Mayo. In
the latter case, an inch or 1½ inches of top sod or 'scraw' was pared off, turned
upside down until partially dry, and then 'footed' or built into small piles until
the sods were completely dry, when they were burned. The land from which
the sods had been removed was made into ridges, and the ashes of the sods
were spread over these. Potatoes were planted on top of the ashes, and covered
with mould from the furrows. The ashes were said to be 'powerful for the
plants', but it was also agreed that burning had 'spoiled a lot of land.'[9]

Unburnt clay and earth, dug elsewhere, were also commonly spread on
cultivated land to increase fertility. The soil was sometimes spread straight on
to the ground, or might first be mixed with farmyard manure.[10] Arthur Young
recorded frequent instances of earth being taken from ditches to form
composts, by both landlords and small farmers. In the Barony of Forth,
County Wexford, for example, he found that, 'They are exceedingly attentive
in getting mould out of the ditches and banks to mix a little dung with it, and
spread it on their land.'[11] The large-scale removal of land from mountains and
bogs was more controversial. The practice was widely reported, however. In
Queen's County (Offaly) in the 1840s, it was reported that, 'In all parts [of the
county] in which it can be obtained, even at a distance of several miles, bog
soil, more commonly called "bog stuff" or manure, is extensively used by large
and small farmers, rich and poor ... I know not what would have become of

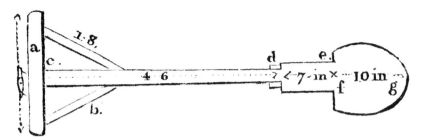

30 A 'skrogoghe' or flachter from County Tyrone, from John McEvoy, *Statistical survey of the county of Tyrone* (Dublin, 1802), p. 51.

the population, or how they would have subsisted without its aid'. In some areas it was alleged that 'bog stuff' was used so much that the land on which it was spread had actually become moory.[12] Contemporaries were more generally worried by the detrimental effects on the land from which the soil had been removed, however. The removal of soil from marginal land was sometimes treated as theft:

> On the ascent from Clogheen … in the Knockmeiledown mountains, we [investigators for the Devon Commission] met several carriers who live by a traffic … which they designate 'stealing mountain'. The stock in trade of this class is a donkey and cart. They derive their means of living from stripping the surface sod, where one can be found, from the unenclosed mountain, and thus accumulate compost, which they store at Clogheen as manure for potatoes for themselves, or to sell to the farmers. This system has been carried on in open day, in spite of the proprietors, for forty years. It is stated that 100 of these marauders have been summoned and fined at one session.[13]

Small-scale removal of soil from hilly areas continued within living memory. In Gaoth Dobhair, County Donegal, the sandy coastal lands of the *machaire* were enriched in this way.

> They would bring down what they called *abar*, that is boggy-stuff, from the hills, and they would mix that and cow dung with the raw sand, and over a period of years it became black and earthy.[14]

ANIMAL MANURE

The organized collection and storage of farmyard manure was widely seen as neglected by agriculturalists. Arthur Young was emphatic:

> In the catalogue of manures, I wish I could add the composts found in well littered farm yards, but there is not any part of husbandry in the kingdom more neglected that this; indeed I have scarce anywhere seen the least vestige of such a convenience as a yard surrounded with offices for the winter shelter, and feeding of cattle.[15]

However, as the quotation suggests, the apparent lack of attention to manure heaps was linked to the lack of planned farm yards and outbuildings for the winter housing of livestock, rather than to ignorance of the importance of dung as a fertilizer. Where animals were housed, animal manure was efficiently collected. This can be seen even in the extreme case of the 'byre dwellings' (fig. 22).[16] As late as 1892, it was reported from west Donegal that, 'cattle in many instances are housed at night at one end of the dayroom [of a dwelling], and the poultry often perch overhead'.[17]

By 1856, the government loans made available to farmers wishing to build planned farm yards included provision for liquid manure tanks.[18] Even in the later nineteenth century, the neglect of manure heaps was still criticized, however. Thomas Baldwin wrote that,

> In small farm districts it too frequently happens that the manure is accumulated in front of the door of the dwelling house, giving rise to the offensive effluvia, which often produce disease. Again, we often see small heaps of manure lying loosely here and there near the house and offices, by which its substance is wasted.[19]

It can be argued, however, that the commonly reported practice of placing the dung heap in front of the dwelling-house door showed the importance attached to animal manure, rather than its neglect, a claim strengthened by the custom, in some areas, of planting ceremonial May-bushes on top of the heap.[20]

SEAWEED AND OTHER FERTILIZERS

In the mid-nineteenth century, the variety of manures used on a large scale increased significantly. Peruvian guano was widely used, as were ground animal bones, and, in the later part of century, the Irish co-operative movement gained some of its most striking successes in the large-scale trade in artificial manures. Throughout the eighteenth and nineteenth centuries, seaweed was widely used as manure, especially for potatoes. Arthur Young claimed that at Westport, County Mayo, the effects of seaweed were so strong that the small farmers would 'not be at the trouble to carry out their own dung hills … A load of wrack is worth, at least, six loads of dung'.[21] In the barony of Tyrawley, he found that, 'The quantity of tillage is very inconsiderable, but what there is, is vastly improved by the use of seaweed. Lands near the sea let at 20*s*. which at two miles, would yield but 14*s*. merely from being too far, as they reckon, to carry the seaweed.'[22] There are many accounts in nineteenth-century writings of the excitement with which people rushed to the shore to collect weed washed ashore during storms:

> Next morning the tempest was still high, and venturing upon the strand, I saw there, as at Valentia, crowds of females busied; and speaking to one, she replied, 'These stawrmy nights, ma'am, blow good luck to the poor; they wash up the say-weed, and that's why you see so many now at work'. The company increased, till I counted more than sixty; and busy, merry work they made of it; running with heavy loads upon their heads, dripping with wet, exultingly throwing them down, and bounding away in glee … 'And are you not cold?' 'Oh no, ma'am the salt say keeps us warm …' 'And how many days must you work in this way before you get a supply?' 'Aw, sometimes not fawrty, but scores of days'. 'And all you have for your labour is the potato?' 'That's all, ma'am, that's all.'[23]

Much more systematic use of seaweed was also common, however. Farms near coasts often had rights to the weed growing on particular strips of beach. In 1837, Binns claimed that in County Kerry, rights were limited to the shore and rocks, but that weed growing under water was free to any one.[24] In other areas, however, 'kelp beds' ran far out into the sea. These were clearly marked off by lines of stones, which are still visible on some aerial photographs (fig. 31). Within these beds, stones were sometimes laid down in sandy areas to encourage the growth of seaweed.[25] In some cases rights to seaweed were sufficiently formalized to be stated in writing.[26] Rights to areas of seaweed persisted into the twentieth century. On the Hornhead estate in County Donegal, for example, it was said that until the estate was sold in the 1930s,

31 Seaweed beds, Carlingford Lough (UFTM L39/2).

tenants knew the limits of the patches belonging to particular farms. Boundaries were not rigidly adhered to when the weed was plentiful, but disputes could arise if a farmer trespassed on to another's patch if the plants were not abundant.[27]

Techniques of cutting growing weed varied. A long pole with a blade at one end was often used, either from a boat, or from rocks at the sea's edge (fig. 32). The late Hugh Paddy Óg Ward of Keadew in County Donegal had clear memories of gathering seaweed during the 1920s. Each farming family had its own area of shore from which seaweed (known locally as 'wrack' or *leathach*)

32 Cutting seaweed, Inis Oir, County Galway, *c*.1900, from Thomas Mason, *The islands of Ireland* (London, 1931).

was cut. Most seaweed was cut at low tide, using hooks. Cutting could take place a considerable distance from the shore, and the weed was brought to shore in boats, usually yawls about twenty feet in length.

> I only mind once being out with my father … [I was about twelve years old] … and it was me that steered the boat. I was on the helm … and they were pulling on two oars … and there wasn't [two inches] … left over the water … only it was quite calm … There was three men on Cruit Island, and they used to help each other out in springtime … cutting the seaweed … So this day … they were coming in with a cargo of wrack, and the little boat they'd with them went on a rock … Well this lad went in to shove her off the rock … And he shoved her off – he was hanging on to the side of her, you see. And one of the boys in the boat … thought the boat was going to capsize, and he got hold of one of the pins … for putting the oar in, and he started hitting him on the knuckles, telling him to let go of the grip. He says, 'Never again would I go out with them boys.[28]

LIME

Agriculturalists were generally agreed that seaweed, either dried or burnt, was an extremely effective fertilizer, but its effects rarely lasted more than one season. The benefits from liming, on the other hand, could last for up to eight or nine years. Apart from limestone rock itself, other materials containing lime were also exploited. Marl clays were an important source. In 1863, marl was described as 'a calcareous earth, of which different specimens show very different amounts of lime in their composition … It is variously designated as clay, stone, and shell marl'.[29] Arthur Young recorded the dredging up of 'shell-marle' in Waterford harbour, and from the Shannon.[30] In 1808, Hely Dutton recorded that

> Astonishing improvements have been made in the neighbourhood of Killaloe, especially in the mountains between that and Broadford, by means of marle, inexhaustible quantities of which may be procured in the Shannon. It is raised by boats and drawn into heaps on the shore, where it generally lies until dry, and at leisure times is drawn to the land; about fifty loads are used to the acre.[31]

Marl clays were found in many parts of Ireland, especially under bogs. In 1839, John Sproule advised that since the marl did not act so quickly on the ground as lime, it should be applied in larger quantities. He particularly recommended its use on 'gravelly, sandy and peaty' soils.[32] Despite the recognition of the value of marl, however, it does not seem to have been exploited as extensively as other sources of lime. In coastal areas, shells and shelly sand were widely used, either being burnt or spread directly on to the land. Shells were identified as an important resource around Lough Foyle as early as 1708,[33] and their use in the north-west has continued until very recently. In 1938, for example, shells were still being burnt for fertilizer on Tory island, County Donegal.[34]

One source of lime, claimed by several writers to be almost uniquely Irish, was limestone gravel. Arthur Young described this as 'a blue gravel, mixed with stones as large as a man's fist, and sometimes with a clay loam; but the whole mass has a very strong effervescence with acid. On uncultivated lands it has the same wonderful effects as lime, and on clay lands, a much greater'.[35] In 1802, Tighe claimed that, 'Before the practice of burning lime came into common use [in County Kilkenny], which is not above seventy years … [limestone gravels and sands] formed the principal manures of the county: for which reason great pits and excavations are to be found from whence they were raised'.[36] In 1824, Dutton also testified to the widespread use of gravel in County Galway: 'Limestone gravel has been formerly used in this county to such an extent, that there is scarcely a field that has not an old gravel pit, and in some places there are several'.[37]

Arthur Young identified the large-scale burning of lime in kilns as a mid-eighteenth century development in several parts of Ireland. He recorded a particularly rapid increase in the number of kilns in the area around Portaferry, County Down: 'There are many kilns ... I was told 35, and that 15 years ago [*c.*1765] there was only one, so much is the improvement of land increasing'.[38] Lord Chief Baron Forster, Young's 'Great Improver', believed that his success in improving land was principally due to the use of lime:

> He had for several years 27 lime-kilns burning stone, which was brought four miles, with culm from Milford Haven. He has 450 cars employed by these kilns ... The stone was quarried by 60 to 80 men ... He spread from 140 to 170 barrels of lime per acre, proportioning the quantity to the mould or clay which the plough turned up. For experiment he tried as far as 300 barrels, and always found that the greater the quantity, the greater the improvement.[39]

The construction of lime kilns was widely discussed by early nineteenth century agriculturalists. In 1802, Tighe recommended that if kilns were not entirely made from limestone, at least the inner surfaces should be given a lining of the rock. This, if necessary, could be replaced every twelve years.[40] Doyle, writing in 1842, recommended that kilns should be built of stone, lined with 'fire-brick'. He also argued that kilns should be narrow, as this led to economy of fuel. In tall, narrow kilns, Doyle claimed, 'more than two thirds of the contents ... [can be] drawn ... every day, whereas ... large circular kilns yield but half their contents in the day, and require considerably more fuel'. However, Doyle also pointed out that many kilns were constructed very crudely, and in many areas, especially in the south and west of Ireland, lime was burnt without using a kiln at all: 'A rude bank thrown up in a temporary way, on the side of any abrupt elevation of ground, answers the purpose of a regularly constructed building'.[41]

The local burning of lime continued on a large scale into the twentieth century. One account from County Tyrone gives a clear description of the process:

One cartload would be a sufficient charge for a kiln. The stone was broken to about three-quarters of an inch mesh. The kiln was charged with alternate layers of peat and stone. The fire was lighted one evening, and during the next two days and one night some one must be constantly in attendance to stoke, first with approximately equal proportions of peat and limestone, and later with peat alone. The peat was added down the chimney. After forty-eight hours the fire was allowed to die, and the lime was raked out through the door.[42]

In 1839, John Sproule listed some of the different ways in which lime was applied to land in Ireland:

1   It may be laid on the surface of land which is in grass, and remain there until the land is ploughed up for tillage …
2   It may be spread upon the surface while plants are growing …
3   It may be, and most frequently is, applied during the season in which the land is in fallow, or in preparation for what are termed fallow crops …
4   It may be mixed with earthy matters, particularly with those containing vegetable remains, and in this case it forms a compost.[43]

All of these methods of applying lime had been recorded by Arthur Young in the late eighteenth century,[44] and continued into the nineteenth century. The decline of 'bare fallowing', however, meant that liming associated with the practice also diminished in importance. Also, although Sproule specifically advised that lime should be mixed only with earth and not dung, mixtures of all three materials were very widely recorded.

Arthur Young, who was usually very critical of Irish cultivation techniques, was unreservedly enthusiastic about the amount of lime being produced from kilns: 'To do the gentlemen of that country justice, they understand this branch of husbandry very well, and practice it with uncommon spirit. Their kilns are the best I have anywhere seen, and great numbers are kept going the whole year through'.[45] Young was aware of the possibility of land being over-limed, but claimed that he had not found this anywhere in Ireland.[46] By the beginning of the nineteenth century, however, some agriculturalists were identifying over-liming as a growing problem. In County Armagh, Charles Coote found lime to be 'prejudicial on light soils'. He alleged that tenant farmers practised liming, 'when they mean to work the soil to the utmost it can produce, or, in other words, to run it out, previously to the expiration of a lease which they do not expect will be renewed'.[47] The changing attitudes to liming in the late eighteenth and early nineteenth centuries can be directly contrasted with those to paring and burning discussed earlier. Paring and burning was at first strongly condemned, but later gained a limited acceptance among agriculturalists as a technique for effectively breaking in some kinds of marginal land. Liming, on the other hand, was at first enthusiastically accepted, but was later treated with more caution. Lavish liming, like paring and burning, was defended as a practice where marginal land was being brought into cultivation. However, Dubordieu's summary of the effects of the over-use of marl and lime echoes contemporary descriptions of the effects of paring and burning: 'From the imprudent use of this manure [marl], the land became exhausted in some years, and fell off in its produce very much … Marle has the same effect as lime … of enriching land, and, like it, the effect of calling out its productive qualities to its destruction.'[48] Similar effects were identified by Greig on the Gosford estates in County Armagh, where upland pastures which had been limed and tilled during the Napoleonic wars were eventually exhausted by the lime's having brought 'every particle of fertility' of the land into action.[49] It has been suggested that overliming was widespread in the early nineteenth century.[50] In 1863, *Purdon's Practical Farmer* stressed how long-term the injurious effects of the practice could be:

> We have sometimes met with cases where it was almost a hopeless task to undertake the improvement of over-limed land, from the state of thorough exhaustion to which it had been reduced; so much so, indeed, that even after being pastured for between thirty and forty years the effects of the overdose were quite visible when the land was broken up. In a case of this kind, the soil when ploughed feels quite puffy and dead under foot; and although oats will braird freely, and seem at first to be very promising, yet in a short time the leaves become brown, and the plants rapidly die away.[51]

CONCLUSION

During the late eighteenth and nineteenth centuries the systematic application of new ideas was gaining ground among large landowners and strong farmers.

There was also an increasing awareness that the techniques associated with a particular mode of cultivation were interlinked and that ignoring this could produce long-term declines in fertility. The integration of cultivation ridges and large-scale drainage schemes suggests that by the mid-nineteenth century some landowners were aware that not all old practices were bad, and that not all new ideas should be followed to extremes. These developments were made explicit in the writings of contemporary Irish agriculturalists. At the other end of the scale, however, the smallest farmers also practised systems of cultivation which could be sensitively balanced between gaining a short term subsistence living, and maintaining the long-term fertility of the land. When, under pressure, these systems were upset, as for example in the over-use of paring or burning or liming, individual small farming families could, and did, face disaster.

CHAPTER FIVE

Spades

THE FIRST TASK in cultivating crops is to till, or turn over, the ground and Irish farming texts from the last three hundred years discuss tillage techniques in great detail. One of the main achievements of early agricultural improvers was to take an explicitly scientific approach to the subject, and this led to major technological advances. However, as with any aspect of Irish agriculture, farmers on the ground also developed their own methods, which despite being slow and labour intensive, could be adjusted to take account of factors such as tiny changes in the land surface and soil quality. This local refinement and ingenuity, which was lost in many large-scale improved systems, is most obvious when we look at spadework.

The wide range of spadework techniques found in Ireland is reflected in the variety of spade designs. During the last three hundred years, there have been more distinctively local types of spade than any other implement (fig. 33). In the southern half of the country, most locally produced spades were made by blacksmiths. During the second half of the eighteenth century, however, specialized mills for manufacturing spades were established. The earliest documented mills were at Ballyronan on the western shores of Lough Neagh, established in the late 1760s, and Monard in County Cork, which probably opened in 1790. Most of the sixty-seven mills eventually in operation were established between 1800 and 1845.[1] Sixty of these were in the north, with the others concentrated around Dublin and Cork. Pattern books used in spade mills categorized spade types by area. As in the naming of dance tunes in traditional music, it seems likely that the same spade type might be known by several different names, but the range of patterns can still seem overwhelming.

As well as the variety in design, the sheer number of implements produced is a clear indication of the importance and complexity of spade tillage in Ireland. In the 1830s, for example, one spade mill in County Tyrone had named patterns for 230 different spades. During the same period, Carvill's mill in Newry was manufacturing 3,500 dozen spades and 57,000 shovels a year.[2]

Some very distinctive styles of spade survived into the recent past. One of the most unusual was the gowl gob (*gabhal gob*), a two-pronged spade used in

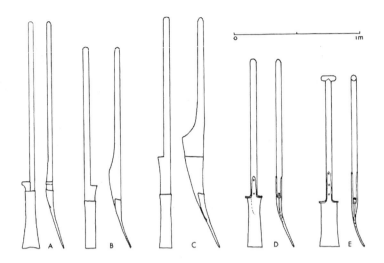

33 (a) Loy made by Scott of Cork; (b) loy made by McMahon of Clones; (c) big loy made by Fisher of Newry; (d) two-sided spade made by McMahon of Clones, with a long shaft, and the 'lift' formed by a crank in the blade; (e) two-sided spade made by McMahon of Clones, with a T-handle on the end of the shaft, and the 'lift' formed by a continuous curve in the blade.

the sandy soils of Belmullet in County Mayo.[3] However, when we begin to categorize spade types in general, a very obvious distinction can be made between one-sided and two-sided spades. All Irish digging spades have foot rests at the top of their blades. Two-sided spades have a foot rest on either side of the shaft. One-sided spades have only one foot rest, set either to the right or left hand side of the shaft. A type of one sided spade known as a 'fack', was used mostly in Counties Wexford, Carlow and other parts of Leinster, survives in museum collections[4] and contemporary illustrations (Plate 1).[5] The fack had a broad blade, and its handle was spliced, which enabled the spadesman to bend it to suit his own working technique.

The most common type of one sided spade is found in the west of Ireland. These spades are known as 'loys' (*láigheanna*). They have narrow blades with open sockets. The shaft and wooden foot rest might be made from one piece of wood, or two separate pieces. In parts of Counties Cavan, Leitrim and Longford a massive spade, known as a 'big loy,' is still sometimes used (fig. 34). The shaft of a big loy is shaped with a 'heel' of wood projecting from the back, to give extra leverage when a sod is being turned. In the past some loys were so big that they were dragged along the ground, rather than carried.[6] In 1832, the

34 John O'Malley making lazybed ridges with a big loy at the
Ulster Folk and Transport Museum, 1997.

agriculturalist Isaac Weld described implements of this type in County
Roscommon. He observed that while the implements were obviously of very
little use for shovelling, they were 'powerful and efficacious' for turning
ground.[7] Big loys had long shafts, made either by a local carpenter or the
farmer himself. In the nineteenth century the shafts on Irish spades in general
were long enough to allow spadesmen to work in an almost upright position.
In the recent past, loys have been most commonly found in the border
counties of Ulster and north Connacht. During the last twenty years, there has
been a revival of the use of the loy in these counties. This is now mainly
confined to competition work, but digging with loys is now one of the most
popular activities organised in ploughing matches in a number of areas.

Many two-sided Irish spades also had long handles. However, in Ulster,
spades often had shorter shafts fitted with a T-shaped handle (fig. 33e). This
may have been a result of improvements to spade design, which will be
discussed below.

In almost all Irish spades, the leverage was increased by an angle in the
blade which made it curve towards the front. The angle or 'lift' might be a
continuous curve, or might be created by a sharp bend or 'crank' in the blade.
In south-west Ireland some blades were bent so that there might be an angle
of as much as 35° to 40° between the lower part of the blade and the shaft.[8]

SPADEWORK TECHNIQUES

The tillage techniques in which spades have been used in Ireland, even within living memory, are as varied as the implements themselves. Their great range is most clearly illustrated by techniques of making cultivation ridges (fig. 35). Growing crops on ridges – raised strips of ground – has been recorded all over Europe, in many parts of Asia, and in the Americas. Cultivation ridges can be made using either ploughs or spades, or both, and as pointed out in an earlier chapter the complexity and rationality of the techniques used provides the clearest evidence for refinement in the history of Irish farming.

As we have seen, ridges were often made to allow crop cultivation in marginal land where soils were poor, thin and badly drained. The most extreme recorded example actually involved creating the soil from which the ridges were to be built. The system, used on the Aran islands, was known as

35 Disused ridges at the Deserted Village, Achill Island.

36 Ridges on 'made' land on the Aran Islands, *c.*1900, from Thomas Mason, *The islands of Ireland* (London, 1931).

'making land' (fig. 36). Thomas Mason outlined the methods used during the early twentieth century for building up soil on the flat areas of bare limestone 'pavement', which are common natural features on the islands.

> Alternative layers of sand and seaweed are placed on the bare rock, any crevices having been filled by broken stones to prevent the 'soil' from disappearing into them, for some of these fissures are ten feet deep; several layers, supplemented if possible by scrapings from the side of the road form the soil in which potatoes and other crops are grown. The ridges are made with a spade and one can hear this implement striking the underlying rock when an Aran man is working his plot.[9]

Where lazy bed ridges are made on grassy ground, the basic technique used is to turn sods over from either side of an untilled central strip, so that they lie inverted in a row along each side of it (fig. 26). (In County Mayo, a sod turned over in this way was known as *feirbín*).[10] These overturned sods build up a central ridge, while the strips of ground on either side, from which the sods have been taken, form the basis of furrows. There are many complex variations in the techniques used throughout Ireland, however. One example recorded

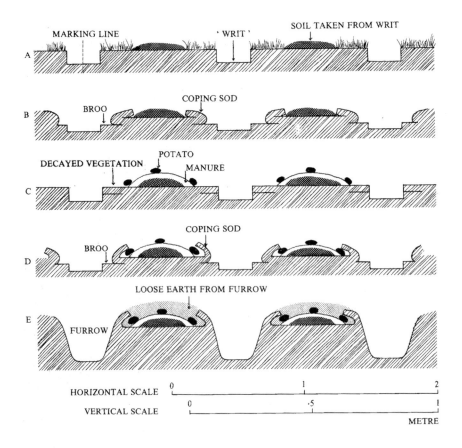

HORIZONTAL SCALE

VERTICAL SCALE

METRE

37 Stages in making spade ridges at Drumkeeran, County Fermanagh.

near Ederney in County Fermanagh during the 1980s shows the sophistication of the system (fig. 37). The highly skilled farmer who demonstrated local ridgemaking methods for the authors was the late Joe Kane of Drumkeeran. The ridges Joe made were normally about three feet (90 cms) wide and 8–14 inches. (20–35 cms) deep. They were made in stages over a number of weeks, usually beginning in February.

The elaborate system used allowed subtle variations to be made according to situation. Joe Kane changed the way he built up ridges, and their overall dimensions, in response to changes in slope, the aspect of the ground being tilled, soil type, the crop being grown, the time of planting, and its place in the annual rotation.

38 Cassie and Paddy McCaughey at Ballymacan, County Tyrone in 1906
(Rose Shaw Collection, UFTM).

Equally striking in these areas of highly developed spadework, was how techniques varied with locality. A distance of only a few miles could mean big changes in the local methods of ridge making. Joe Kane related how some he had gone with some neighbours to a farm only twenty-five miles away, to make potato ridges for a man near the town of Garrison. The men were amused, and confused, by both the type of spade used in this area, and the narrower ridges they were expected to make.[11]

Shovels were often used along with spades (fig. 38). Their importance can be seen from the number manufactured by mills such as Carville's, listed above. Shovels were particularly important in areas where spade blades were very long and narrow, good for leverage, but not for lifting loose earth. Estyn Evans found that Irish shovels, like traditional spades, were long handled. He classified them by the shape of their blades into two types, round-pointed and square-pointed.[12] Most surviving shovels have metal blades, but in the mid-nineteenth century, Sir William Wilde claimed that 'until very recently the shovel common in Galway and other parts of the west of Ireland, was composed of wood, shod with iron round the edge for about two inches in depth; it was usually made of sallow.'[13]

IMPROVED SPADES

Agriculturalists had very different views on the efficiency of Irish spades, especially those of older design. In 1808, for example, Hely Dutton condemned loys and facks as 'inconvenient heavy tools … most unfit for cutting in bog, or moving loose earth'[14] compared to two-sided spades. Some writers also criticised the upright position in which many Irish spadesmen worked.[15] However, most experts agreed that loys were in fact excellent implements for making lazybeds in heavy soil. A comment on loys used in Counties Donegal and Leitrim in 1802, is typical. 'This sort of machine is admirably well adapted to the weight and tenacity of the soil. A broad short spade, pushed into this ponderous gluey stuff, must remain there, as in a locked vice, whereas this narrow one … [with] the long handle answering as a lever and [the heel on] its back … as a fulcrum … very much facilitates the labour'.[16]

Some experts argued that the loy was in fact superior to the two-sided spade. It was claimed that a team of twelve men could turn an acre of ground a day using loys, twice the speed at which an equivalent team of men using two sided spades could work.[17]

By the 1870s, it was accepted that, within the restrictions imposed by high labour inputs and relatively slow rate of working, spadework of all kinds was a very effective way to till ground. A model farm of five-and-a-half acres at the Glasnevin Agricultural College in Dublin, where tillage was all done with spades, was claimed to be the 'one of the best tilled pieces of land in Ireland.'[18]

SPADE LABOUR.

Farming Implement Factory,
Bridgefoot-st., Dublin.

SIR—We have the pleasure to send you herewith an Engraving of the Improved Spade, recommended by the Royal Agricultural Society of Ireland, as the Implement best calculated to promote the successful adoption of Spade Husbandry. The blade is sufficiently long (16 inches) to accomplish the required depth; it is parallel in the sides, and but 5 inches wide; so that (once entered) it is easily forced into the earth, its area or surface incurring a much smaller amount of friction than that of the ordinary Spade hitherto in use.

To compensate for the loss of strength which its long, narrow form would seem to imply, instead of having one Rib down the centre, as in the old Spade, it is furnished with *two*; or so strengthened on the edges (particularly at the weak point, B), as to render it secure against the danger of its yielding there. This arrangement also gives it what is technically called, a good "rise and gether." Its entire length is 4 feet 8 inches, and its price, 27s. per dozen.

As the desired length of handle may vary with circumstances, we also make them with plain, long handles, which may be *cut* to suit the workman's *height*, and taste, at 25s. a dozen.

We are your obedient servants,

HENRY SHERIDAN and CO.

The Improved Field Spade for Manual Labour.

THE SUBSCRIBER begs to direct the attention of Landlords, Farmers, and Agriculturists to the IMPROVED FIELD SPADE so much approved of by the Royal Agricultural Improvement Society of Ireland, and the Royal Dublin Society, as being the best implement for the prosecution of Spade Husbandry, and recommended by the Society's Practical Instructor, Mr. Quinn, for its cheapness and great advantage in agriculture.

Persons desirous of obtaining the Improved Spade, can be supplied on application at the Manufacturer's sole Agents' in Dublin, Messrs. WM. DRUMMOND and SONS, Seedsmen, 58, Dawson-street, and Mr. JOHN MAGUIRE's, 10, Dawson-street; or in NEWRY, at the maker's Foundry, who is the original manufacturer of this useful implement.

FRANCIS CARVILL.

Newry, Jan. 16, 1850.

☞ Specimens of this implement can be seen at the office of the Secretary to the Royal Agricultural Improvement Society of Ireland, 41, Upper Sackville-street, or at the Royal Dublin Society's Agricultural Museum, Kildare-street, and its advantages explained by the Curator.

CASH PRICE—

Men's Spades, 20s. per dozen.........1s. 10d. each.
Boys' Spades, 18s. per dozen.........1s. 8d. each.

39 Henry Sheridan's and Francis Carvill's advertisement for improved spades, in the *Irish Farmer's Gazette*, vol. 9 (Dublin, 1850).

Despite claims made for loys, however, it was a new type of two-sided spade that provoked most debate. In 1850, the Royal Agricultural Improvement Society appointed thirty six Practical Instructors, based mostly in western counties, who taught improved spade tillage techniques associated with systems of green cropping. This initiative led to a fiercely competitive advertising campaign between two major spade manufacturers, Henry Sheridan of Dublin and Francis Carvill of Newry, both seeking the market for an improved design of spade. Evidence suggests that it was in fact northern manufacturer Carvill who developed the new implement, probably around 1830 (fig. 39).[19] Wherever the spade originated, its development marked a move away from the production of local spade types suited to local tillage practices, to the mass production of a standardized implement suitable for deep tillage and use in large scale, systematic cultivation systems using spade labour.

In the late 1840s, the 'improved spade from the north of Ireland' was reported in use as far south as the Fermoy Union in County Cork and the Lismore Union in County Waterford.[20] One of the most dedicated champions of spadework, Alexander Yule, a Scot who was employed as land steward to the Anglican Primate's Palace Farm in Armagh, described the spade as an 'Armagh' type, which was recommended by the Royal Agricultural Society.'[21] It was a two-sided spade, which kept some older Irish features, in particular the relatively long blade, slightly curved to increase leverage. Its introduction to southern parts of Ireland was often associated with drainage projects, where spades were used on newly ploughed ground to break up subsoil, and the new spade blades had to be long enough for use in this deep tillage. When the spades were introduced to the Fermoy Union in 1849, local labourers were reluctant to use them at first, but this soon changed to enthusiasm, the men claiming that they could do twice as much work with the improved spades as they could have done using their old facks.[22] The complex construction of the new spades, and northern spades in general, probably put the production of the large numbers of implements required beyond the scope of local smiths, and it may have been this which led to the concentration of specialist spade mills in the north.

Improved spades were sometimes known as 'garden' spades, and were used in intensive horticulture, and in the technique known in standard British works as 'trenching'. This technique was designed to deepen productive levels of soil by mixing upper and lower levels, and enriching the latter with fertilizers. It involved digging over the ground surface in a series of parallel trenches, the top layers of one trench being placed in the bottom of the one preceding it. The intention was to mix upper and lower layers of soil together, bringing subsoil to the surface, where it could be broken up, aerated and manured. By the 1890s, however, the value of trenching was questioned. It had become clear that burying topsoil could lead to the destruction of small organisms which lived in it and contributed to its fertility. It was also recommended that subsoils which were composed of cold clays should not be brought to the surface, as these would increase drainage problems in upper levels of the soil.[23] The trenching technique was seen as most effective in clay or peaty soils, which lay over gravel or sand.

In many parts of Ireland, the introduction of improved spades went along with the development of a programme aimed at establishing standardized techniques throughout the country, taught by the Practical Instructors mentioned above. Competitions were one way in which the standardisation could be pushed forward. In 1849, The Royal Agricultural Improvement Society attempted to co-ordinate its encouragement of digging matches. In 1850, the *Irish Farmer's Gazette* called for a standardized format for recording these, including soil quality, previous use of the ground, the depth dug, the sort of spade used and its blade length, whether the ground was level or sloping, the crop grown and the general size of sod turned, including a note on whether it had been 'carefully and completely turned upside down.'[24] There was a lot of early evidence that the movement was succeeding. The Wexford Union Agricultural Society's digging match in February 1850, for example, was reported as a evidence that the new spades and associated techniques were being enthusiastically adopted by farm labourers.

> Thirty-five as able and athletic fellows as ever we beheld entered on the task; and on the word being given, dashed at their work like men determined to 'do or die'. It was evident, from the adroit and workman-like manner in which most of them handled the iron spade, that its superiority is at length being felt by the labouring population.[25]

PLOUGHS VERSUS SPADES

Just before the Great Famine of the 1840s, agricultural labourers and their families probably made up more than a quarter of the Irish population.[26] This vast number of people (over 2.25 million) needed some way of making a living, but the supply of labour was much greater than the demand, and many employers were tempted to offer wages at 'the lowest amount which will support life'. Long term developments in rural society, such as a falling birth rate and increased emigration may have altered this situation, but these slow changes were brutally disrupted. Between 1845 and 1849 the potato crop, on which 3.3 million Irish people depended as their main source of food, failed year after year. At times, especially in the darkest year of 1847 when the deaths from disease and hunger were at their worst, even reluctant government officials showed awareness that new and radical emergency measures must be considered. Soup kitchens and public works were the most common relief measures set up by the state. At this time, one energetic group of campaigners also tried to help by arguing that spade cultivation, even on large farms, would be an effective and humane way to deal with the crisis, not only to relieve the horrific suffering of the poor, but to take pressure off the collapsing workhouse system and ensure the maintenance of law and order.[27]

Alexander Yule, Land Steward to the Anglican Primate of Ireland, suggested that the relationship between the labourer and his spade was part of the natural order of things.

> The system of agriculture must be changed to *make* employment, and thereby save taxation, and keep down pauperism, want, misery and crime … If by bringing [manual husbandry] … into general use, we can employ *all*, and reconstruct our social system on more agreeable terms … we shall be … teaching the peasantry better lessons than they are likely to receive at either a workhouse or a meal depot … Deprived of the *spade* which Nature intended they should use, they seize the musket or the blunderbuss … My object is to disarm the peasant but to employ him.[28]

Some campaigners made very ambitious claims for spades as efficient tillage implements. In a public lecture at the Rotunda in Dublin in 1848, for example, Mr N. Niven claimed that while

> Much has been said, and a great deal written, as regards the application of the plough in husbandry; and although, no doubt, the improvements in the mechanism of this most important implement, have been, of late years, very great, still not withstanding all, I do conceive that its efficiency, as an implement of tillage can never be compared with that of the spade.[29]

One of the main advantages claimed for spades over ploughs, was that spades broke up layers of subsoil, while ploughs tended to compact them.

> [The] plough may be said to be very inconsistent, inasmuch as while the plough is partially pulverising the soil, the horses' feet are consolidating [it], and in consequence the whole should at once be abandoned, and give place to the poor, cheap, neglected spade, which is superior in every respect.[30]

Campaigners for the spade also pointed out that the cost of buying and maintaining ploughs, harness, and especially horses, would all be saved if spade labour was used instead. It was claimed that savings of up to £2 an acre might be made by using the cheaper, superior method of spade labour. Other commentators, however, pointed out that a lot of these savings were only made by giving unacceptably low rates of pay to spadesmen.

The argument became especially heated in the later phases of the Famine. Advocates of the plough were as assertive as campaigners for spadework. A Mr James Hunter of Glenville, County Cork, for example, was dismissive of the notion that spades could replace ploughs. Discussing his own project of tilling 350 acres using only horses, he commented,

> The cuckoo song of giving more to the unemployed will be brought in; but against this I state, the money retrenched by the excellence of the plough is otherwise expended in a progressive direction, in making more work for *her*, not to be conquered – *the plough*.
>
> This is the way to make progress, and there is no other way for Ireland's recovery. Money may, for a time, bolster up a spade-mania-delusion, but nothing but disappointment must be the certain result.[31]

The bitter response by the newspaper in which Mr Hunter had stated his views, shows the intense emotion which these apparently prosaic issues aroused during the Famine.

> It is evident that [Mr Hunter's] … plough and horse system … has removed and thrown out of employment numbers of our population. We look upon Mr Hunter and his system as the real exterminators, the transporters, and alienators of the strength and sinews of the land to other climes – nay to the grave; for if we had two or three Mr Hunters located in every parish in Ireland, the country, of course would become depopulated.[32]

The debate continued into 1850, but the heat went out of it very quickly. In 1851, the Royal Agricultural Improvement Society abolished the Practical Instructor scheme, which had been central to the movement to promote improved spade techniques, and the organized drive to promote spade husbandry declined very quickly. Discussion continued, to some extent. In the early twentieth century, some Irish agriculturalists still argued that where labour was plentiful, the use of spades had ensured the most productive tillage system for crop production in the nineteenth century. In 1920, officials of the Department of Agriculture and Technical Instruction cited the transition from spade to predominantly plough cultivation as a contributing factor to the prolonged drop in crop yields after the early 1850s.[33] During most of the twentieth century, spadework continued to be important on small western farms in particular, and even in the new millennium, older men in these areas can still demonstrate complex spadework techniques. However, the large-scale use of spades was not widely advocated after the mid-nineteenth century. The long term movement away from crop cultivation to livestock farming, the post-famine decline in numbers of agricultural labourers, and the general increase in farm size, meant that labour intensive techniques like spade work increasingly gave way to horse drawn techniques, such as ploughing.

CHAPTER SIX

Ploughs

AGRICULTURAL IMPROVERS OF the eighteenth and nineteenth centuries were more interested in horse operated implements than in hand tools, and ploughs were more widely discussed than any other implement. Ploughs have been used in Ireland for thousands of years. For most of this time, Irish ploughs changed in ways that can be paralleled in the rest of Europe, but they also had distinctive local characteristics. During the early eighteenth century, however, a long period of international experimentation with plough design began, and commentators often compared the new implements produced to the ploughs which were in common use (Plate 2). 'Common' Irish ploughs were generally described very critically. In 1844, the great Irish agricultural writer, Martin Doyle, commented that a field turned by one local type of plough looked 'as if a drove of swine had been moiling [ruining] it.'[1] However, there were several different types of 'common' plough, and recent research suggests that some of these older implements performed very well in particular cultivation systems.

Internationally, the simplest type of plough is known as an *ard*. The main working part of these implements is a point of wood or metal, which rips up the ground as it is pulled through it. Prehistoric ards have been found in Ireland, and the use of ard-like implements was recorded in the nineteenth and twentieth centuries.

In Fermanagh during the 1920s, these implements were known as 'oul' stick ploughs', while in nineteenth-century Donegal, they were said to be constructed from 'a crooked stick with a second one grafted on to it, to give two handles, and the point of the stick armed with a piece of iron.'[2] Recent evidence collected during fieldwork in County Roscommon, shows that a loy spade could sometimes be used as an ard, harnessed with chains to a donkey. In 2005, one County Longford farmer, Mr Michael McKeown, demonstrated the use of a loy harnessed in this way. Here, the loy was used to loosen soil in the bottom of furrows between ridges, which was then shovelled on to the ridges to cover seed (fig. 40a, b).

A neighbour of Michael McKeown's, Mr John Madden, said that in the past, the loy/plough could also be used to lever over sods that had been marked out using the spade.[3]

40 Michael McKeown of Aughnacliffe, County Longford, demonstrating: (a) the use of a loy to turn a sod; (b) the use of a loy harnessed to a donkey, to make lazybed ridges.

THE OLD IRISH LONG-BEAMED PLOUGH

Since medieval times, most ploughs used in western Europe have had three main working parts (fig. 41). The coulter cuts a vertical slice through the ground, the share (or sock) makes a horizontal cut below the vertical slice

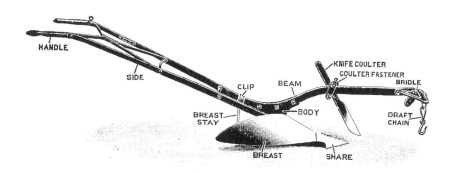

41 Swing plough, showing working parts.

made by the coulter, and the breast or mouldboard – a wooden board or metal plate attached to the side of the plough – overturns the long strip of earth cut by the action of the share and coulter out of the ground (the furrow slice). The shape of the mouldboard controls how effectively the furrow slice is turned over. The most frequently discussed 'common' ploughs were of two basic types; the Old Irish Long-Beamed Plough, and the Light Irish Plough.[4]

The Old Irish long-beamed plough had the working parts just listed, but its mouldboard was flat rather than curved, as in almost all improved ploughs. Ploughs from Counties Meath, Derry and Kilkenny were described in detail in the early 1800s, and there is at least one fairly complete surviving implement, from Moira in County Down (fig. 42i, ii, iii).[5] However, we do not know of any complete specimen of an old Irish long-beamed plough, and we have to rely on the written descriptions, and a few good illustrations, to work out details of their construction and use.

Shares of Irish long beamed ploughs were set at an oblique angle, rather than fixed horizontal to the ground. Along with the flat mouldboard, this meant that the plough turned loose furrow slices that stood up almost vertically, something that was severely criticized by improvers, who preferred ploughs which turned furrow slices over so that they would lie neatly inverted on the ground. The long beamed ploughs were also criticized for their size and weight. Three people were usually required to operate them successfully. A ploughman controlled the plough using its handles, a driver walked in front of the team of animals pulling the plough, to guide them, and sometimes a man or boy would be required to lean on the plough's beam, or to press down on it with a stick, to keep it in the ground.

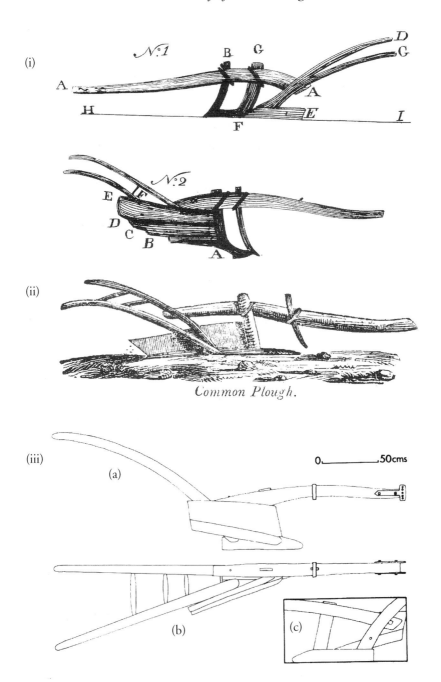

42 Common Irish ploughs: (i) County Meath, from R. Thompson, *Statistical survey of the county of Meath* (Dublin, 1802), p. 108; (ii) County Kilkenny, taken from W. Tighe, *Statistical observations relative to the county of Kilkenny* (Dublin, 1802), p. 294; (iii) Moira, County Down. Inset shows detail of land side.

Improvers also criticized the number of animals required to pull the plough. Both horses and oxen were used for ploughing, although ox teams were much less common, and were mostly used on the farms of landlords or gentlemen farmers. Four, or even six animals were needed, depending on the condition of the ground being ploughed, and the sort of work being carried out. Four animals could be used for the initial work of breaking up ground in winter and spring, but six might be required for the heavier work of ploughing summer fallows.[6]

Fallowing was an element of many crop rotation systems. There were two main types of fallowing used in early nineteenth-century Ireland. In 'green' fallowing, a root crop such as turnips or potatoes was planted, to break up and clean ground. Summer or 'naked' fallowing, on the other hand, involved repeated ploughing of ground during the year. This system usually involved cross ploughing, a secondary process, known in Ireland as *gorail*, or 'gorrowing', where a field was ploughed at right angles to the direction of the initial line of ploughing. The technique allowed the depth of productive soil to be increased, but it meant that the plough had to be put much deeper into the ground, which is why a larger plough team was needed. Summer fallowing was often used to prepare land for wheat cultivation, and the big, heavy long-beamed plough, with its draught provided by up to six animals, may have been most common in the main wheat growing areas of Ireland, which were mostly found in Leinster.[7]

LIGHT PLOUGHS

Light 'common' ploughs seem to have been associated with areas of easily worked soil. In 1854, in north County Donegal, the local plough type was said to be so light that a boy of eighteen could carry one on his shoulder. The ploughs were used to prepare ground for potatoes and grain in areas where a team of two, or at most three, horses was sufficient to pull the plough. A similar plough was observed in use on light soils in County Sligo.[8]

Light ploughs were also sometimes used in a secondary role, along with heavier long-beamed ploughs. The light plough might be used to loosen earth in the furrows between ridges that had been built up using a heavy plough, the loosened earth then being shovelled on top of the adjoining ridges. Light ploughs might also be used to cover potato or cereal seed, where it was thought that harrows would not cover the seed deeply enough.

43 Eighteenth-century light wooden plough from Aughnish,
County Galway, preserved in Power's Distillery, Dublin.

We know of one surviving light wooden plough, which is said to date from
around 1780. It originally came from Aughnish, County Galway, and has been
preserved in Power's Distillery in Dublin (fig. 43). It has some 'modern'
features, such as a short beam and relatively long handles, and a convex curved
mouldboard. It may be a direct ancestor of wooden ploughs which were made
and used in the same area of south County Galway and east County Clare
until well into the twentieth century.

IMPROVED PLOUGHS

By the 1750s, members of recently formed Irish farming societies were
sufficiently interested in new plough types to have a classic British text,
Thomas Hale's *Compleat Body of Husbandry* reprinted in Dublin.[9] This work
described and illustrated a number of experimental ploughs, which had wheels
fitted at the front.

Wheels were commonly fitted to ploughs in lowland Europe and England.
They were intended to make the plough run over the ground more smoothly.
On some ploughs, adjustments could be made to the relative height and
distance at which wheels were set, and this helped the ploughman to control

44 Medal of the Farming Society of Ireland, showing a
Scottish swing plough pulled by two oxen, *c.*1815.

the depth and width of the furrows turned. In 1767, an Englishman, John
Wynn Baker, manufactured eleven types of plough in his factory at Cellbridge,
County Kildare.[10] Only two of these ploughs had wheels, but around the same
time some landlords introduced wheel ploughs to their Irish estates. These
were older plough types, associated with particular regions of England,
including Norfolk, Suffolk and Kent. In the 1770s, Lord Shannon went so far
as to bring a plough and ploughman from France. This was part of his project
to introduce oxen as working animals on to his estate farm. (His workers
formed a combination to oppose the changes, until he offered higher wages to
men prepared to learn the new skills required.) Arthur Young, who noted these
landlord experiments, also found that wheel ploughs were used on a small
Palatine colony in County Kerry, while in their colony in County Limerick the
Palatines had introduced ploughs fitted with hoppers through which seed fell
on to the newly-turned ground.[11]

Most of these early, tentative experiments show Irish agricultural improvers
looking to England for new implements and techniques, but by the early
nineteenth century there was widespread agreement that wheeless 'swing'
ploughs of Scottish design were better suited to conditions in most of Ireland

(fig. 44). In the Barony of Imokilly, County Cork, in 1810, for example, a 'good many' Scottish ploughs had been introduced, but only a few English ploughs.[12]

The most influential Scottish ploughwright was James Small, who attempted to design ploughs based on scientific principles. Small's main achievement was to develop mathematical methods for calculating the best curvature for plough mouldboards. This determined how effectively the plough could turn furrow slices over after they had been cut by the share and coulter. Small argued that, 'The back of the sock [share] and the mouldboard should make one continued fair surface, without any interruption or sudden change. The twist therefor, must begin from nothing at the point of the sock, and the sock and the mouldboard must be formed by the very same rule.'[13] His early ploughs had wooden mouldboards, but after 1780 he began to have them cast in iron. His plough designs were developed by other Scottish inventors, such as Wilkie of Uddingston, and by the early 1800s few experts had anything good to say about 'common' Irish ploughs when comparing them to the new Scottish models. The new ploughs required teams of only two horses or oxen for draught, and they could be efficiently operated by one ploughman. It was also generally agreed that they turned over sods much more effectively.

By the 1820, most ploughs used in Ireland were probably of Scottish design. Some Irish farmers were slow in adopting the new ploughs, however, and this provoked impatience from improvers. In 1802, the agriculturalist Hely Dutton dismissed the caution of farmers in County Dublin as arising from ignorance and prejudice. '[They] say it will not do for this country; that it kills the horses etc., etc.'[14] He did, however, recognise that some of the concerns were reasonable. To work efficiently, the irons (the coulter and share) on the new ploughs had to be set properly in relation to one another. This meant that new skills were required of manufacturers and blacksmiths, as well as ploughmen, before the ploughs could be operated properly. Other concerns included the initial expense of buying a plough, and the loss of employment for hired hands who might have expected work when the older, more labour intensive ploughs were used. This loss of employment was the subject of a piece of doggerel in the *Dublin Penny Journal* in 1833.

> An Irish ploughman, with much toil and pain,
> Had worked a light Scotch plough against the grain,

For seven long years reluctant in Fingal,
Being asked, 'pray Paddy how d'ye like it now?'
Exclaimed, 'Ochone, give me the Irish plough,
It is the plough for Ireland after all!'
'Why, Paddy, you're a most ungrateful rap!
The Scotch with ease works nearly double.'
'Ah sure enough, and saves a world of trouble,
but still the Irish is the plough for crap.'
'Nay, Paddy, that you know is not the case.'
'Why sure enough your honor's craps increase'
'Then what objection can you have, you oaf?'
'Why then, I'll tell your honor's honor why
Before your honor laid the ould plough by,
We always had a mighty bigger loaf.'[15]

The extra labour required to use the older ploughs had often come from the ploughman's family.

Old ploughmen are very much averse to the Scotch plough and they therefore throw every obstacle in its way; ... perhaps ... [the ploughman's] son has been the driver upon the old system, in which case little is to be expected from the father's exertions, as the greater the degree of perfection, to which he brings the Scotch plough, the farther his son be from employment.[16]

One solution suggested to this problem was that farmers should pay ploughmen, first their own wages, but also half of the driver's wages.[17] This suggests how determined agriculturalists were that the new Scottish ploughs should be used. Even the most enthusiastic supporters of the new ploughs, however, agreed that in some situations they did not perform as well as 'common' Irish ploughs. Richard Thompson, for example, admitted, 'I am, and have been, since I first saw the Scotch plough, a very great advocate for its being generally adopted, yet I confess that in this country there are many which it will not answer for the summer's gorrowing [i.e., cross ploughing summer fallows]'.[18] As late as 1819, William Shaw Mason reported that in County Kilkenny, although

Some of the more respectable farmers have introduced Scotch ploughs … they are not suited to wet grounds, which are so stiff and adhesive, that the sod requires to be raised as much as possible, in order that the air may get through it, to dry and pulverise it. The Scotch plough is thought to lay the sod too flat, while the Irish plough raises it, and breaks it more; … [while] the superiority of the Scotch plough in dry grounds is fully acknowledged, a dislike to it … in the former instance may have some foundation in reason. However when, in the advancing improvement of agriculture, those grounds were drained … pulverisation will be effected by the Scotch as well as the Irish plough.[19]

Reluctance to use Scottish ploughs, therefore, was not simply due to unreasonable conservatism. New skills had to be learned by ploughmen and plough makers, the initial cost of buying the plough was greater, ploughmen were concerned at possible job losses arising from the new technology, and the new ploughs only performed at their best on well-drained land, and in systems of agriculture where summer fallowing was not practised. Hely Dutton recognized how easily new implements and techniques could be discredited:

We may continue to import Scotch swing, and English wheel-ploughs, and other implements of utility and whim, but unless we import along with them the best practices of each country, and steadily pursue them in opposition to the old school stewards, it will only serve to bring them in disrepute with those who are but too ready to catch at any opportunity to decry practices they do not understand.[20]

There were several attempts to improve Irish ploughs, but the adoption of Scottish swing ploughs did continue. In 1812, the Farming Society of Ireland employed a champion ploughman, William Kippie, to instruct farmers in the use of Scottish ploughs. Kippie would 'attend any farmer who wishes to employ him 2s. 6d. per day with his diet and lodging'.[21] The enthusiasm of farming societies, and their success in persuading Irish farmers to start using the new ploughs, can be traced through accounts of ploughing matches reported in the *Munster Farmer's Journal.*

In 1812, the Cork Institution decided that at its ploughing match for the next year there should be two classes of premium: one for the ploughmen of 'gentlemen and farmers', and the other for 'working farmers'. Irish ploughs could only be used by the second category.[22] Ploughs of Scottish design, some made locally, were given as prizes. In a ploughing match at Fermoy, County Cork, in 1813, separate prizes were given for Irish and Scottish ploughs,[23] but by 1816 the Scottish ploughs were becoming dominant. At a ploughing match at Kinalea, County Cork, on 13 March, 21 ploughs were entered, 20 of them Scottish design. The report on the match concluded that

> The rapid improvements in ploughing … and the very general adoption of Scotch ploughs which have been introduced into this district within the last few years, afforded a most satisfactory evidence of the utility of ploughing matches – and the great object for which they were instituted having been achieved, the members have resolved to limit the number such contests, to one annually.[24]

Even those County Cork farmers who had not actually bought a Scottish plough were said to have improved their own implements, and learn to plough with two horses and without a driver.[25]

Small, locally designed wooden ploughs survived, however, sometimes as subsidiary implements to the metal Scottish ploughs. In 1837, for example, Jonathan Binns found that in the west of County Waterford 'Previously to sowing grain, the land is ploughed with the Scotch plough, and well harrowed; then a small wooden plough without iron in the mouldboard is used for the purpose of ploughing in the seed. Under this process, the ground, not being left smooth, is less liable to become hard and stiff.'[26]

LATER IMPROVED HORSE DRAWN PLOUGHS

By the 1850s, all-metal swing ploughs were used throughout every commercial farming region in Ireland. Mouldboards for these ploughs had to be cast, and small foundries produced standardised ploughs. The best known foundries were established in the east of Ireland, and around Cork. They included Grays of Belfast, Paul and Vincent of Dublin, and Ritchies of Ardee.[27]

Some local foundries, such as Kennedys of Coleraine, produced plough parts, which were then assembled by local blacksmiths. Blacksmiths acted as a link between foundries and farmers, and had to be knowledgeable about both the industrial production of the implements, and how they were used. Many farmers could adjust the position and angle of the plough coulter for

A　　　　　　　　　B　　　　　　　　　C

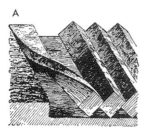 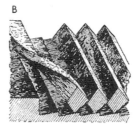 

45 Three types of furrow slice produced by ploughing: (a) rectangular work; (b) high cut work; (c) broken work, taken from R.P. Wright (ed.), *The standard cyclopaedia of modern agriculture and rural economy*, 12 vols (London, 1908), vol. 9, p. 252.

themselves, but the fine work of adjusting the wing (or feather) on the plough share usually required the skill of a smith. Some blacksmiths tested the angle of the wing using a template.[28]

Very small changes to the angle at which the wing was set (by hammering it on an anvil), could significantly change the shape of the furrow slice turned. If the share was hammered very flat, and the plough was fitted with a concave mouldboard, a rectangular section furrow slice was turned (fig. 45a). To turn furrows with sharper, high crested profiles (fig. 45b), a convex mouldboard was required, and the wing of the share had to be raised to a very precise angle.

Sharp-edged furrow slices made an excellent seed bed for crops such as grain. When the seed was broadcast over the ploughed ground, it tended to fall in rows between the long straight furrow crests, and when the field was harrowed, soil on the furrow crests was easily knocked off to cover the seed. In carefully ploughed fields, the pattern of sowing in long straight, equidistant rows was not unlike that produced by seed drills.

Horse-drawn swing ploughs remained common in Ireland until well into the 1950s. This was especially the case on small farms in hilly areas. A swing plough required a lot of skill from the ploughman using it, because he could not rely on wheels at the front of the plough to help control the depth and width of furrows being turned. However, wheelless ploughs could be made to respond very sensitively to changes in slope, or even bumps in a field. By pressing down on the handles, or lifting them even slightly, the ploughman could adjust the depth at which the plough went through the ground. The width of furrow slice cut by the plough could also be changed, by sideways pressure on the handles. This meant that when going up hill, the ploughman could make the plough cut a narrower, shallower furrow slice, so easing the draught on the plough team. Swing ploughs could also be eased over large stones, without bouncing off them, so avoiding breakages, particularly of ploughshares.

On level ground, however, ploughs with wheels attached to the front had the great advantage that the ploughman could fix the depth and width of the furrow slice to be turned (Plate 3). The wheels at the front were adjusted so that one wheel ran in the bottom of the furrow, and the other on top of the unturned grass.

The relative heights at which the wheels were set, fixed the depth of furrow slice turned. The horizontal spacing of the wheels, fixed the furrow width. This meant that with a well-trained team, the ploughman could produce neat, accurate ploughing, with relatively little effort. Wheel ploughs operated best on level, well-drained ground, and not surprisingly, the biggest plough manufacturers were based near Ireland's richest tillage regions of Leinster and north Munster. Wexford town had several foundries. The largest of these, Pierce of Wexford, and the Wexford Engineering Works, produced wheel ploughs based on English models, which became more and more standardized as the international exchange of agricultural technology increased. The most famous of these was Ransome's Universal Body Plough, which was built up of separate pieces, bolted together. Farmers could unbolt broken or worn-out parts from the plough, and replace them with parts bought from a local supplier. By 1950, some ploughs manufactured in Wexford could be fitted with parts made by a number of British firms, including Ransomes, Howard, Hornsby and Sellar, and the American firm of Oliver. The latter firm, the Oliver Chilled Plough Works of Indiana, made ploughs which were fitted with revolutionary 'digging' mouldboards. These turned a wide, broken backed furrow slice (fig. 45c) that was particularly useful in preparing ground for drill cultivation, which will be described in the next chapter. The American ploughs allowed the soil to be broken up to a sufficient depth to allow raised drills to be set up for planting potatoes, and reduced to a fine enough tilth for grain crops to be planted easily with a horse-drawn seed drill.

PLOUGHING TEAMS

Up until the First World War, there was an ongoing, if sporadic, debate in Irish agricultural writings as to whether oxen were as good, or better, working animals as horses (fig. 3).[29] By the 1900s, however, this was a rather esoteric

46 An inaccurate, anti-Irish caricature, showing ploughing by the tail, 1808, from Edward Dubois, *My pocket book; or, Hints for 'A ryghte merrie and conceitede' tour* (London, 1808), p. 82.

discussion, as most Irish farmers had in fact been relying on horse power for several hundred years. The introduction of improved all-metal ploughs meant that the typical nineteenth-century plough team was made up of two horses, harnessed abreast.

The most notorious system of harnessing recorded in Ireland, was known as ploughing by the tail (fig. 46). This had provoked contempt from colonial administrators as early as the seventeenth century, and the Irish Parliament passed a law forbidding it in 1634. The practice was sporadically recorded in Ireland and western Scotland until well into the nineteenth century – in Erris, County Mayo, as late as 1841.[30] (The authors were told of an instance in north County Down as late as the 1930s, when a local farmer (seen as both mad and bad) had tied his horses' tails to a turnip planter, when sowing seed.) There have been attempts to rationalize the practice, but it is not really defensible, as a cheap, efficient, and humane alternative was widely known and used. Farmers could make their own harness collars and backbands from plaited twisted straw (*súgán*) (fig. 47).

Trace ropes, which were required to be strong enough to take the strain of horses pulling a plough, could be made by twisting strips of bog fir into very

47 Mick and John McHugh with a súgán collar, at Hornhead,
County Donegal, in 1982.

strong ropes (*gaid*). Some older farmers still remembered how to make this
harness as recently as the 1980s,[31] but during the nineteenth century leather
harness became standardized, with very few regional differences in design.
There were two main types of horse collar made in Ireland; one open ended,
which was slipped around the horse's neck and strapped at the top, and the
other closed, put over the horse's head upside down, and then turned around.

PLOUGHING TECHNIQUES

During the eighteenth and early nineteenth centuries, ploughs, like spades,
were used mostly for building up cultivation ridges. Techniques of making
ridges varied widely throughout Ireland, as also did their dimensions when
complete. In 1774, for example, Hyndman discussed the cultivation of flax on
ridges ranging from three feet to twenty-one feet in width.[32] Very broad ridges
were recorded in County Kilkenny in 1802, where some 'gentlemen' farmers
practised a technique of ploughing 'in balk' for corn crops. The ridges made
were up to sixteen yards wide.[33] In the mid-nineteenth century, however, the
agriculturalist Martin Doyle claimed that the common Irish practice was to
make 'extremely narrow' ridges, about four feet in width.[34] The widths of
furrows, or trenches, separating ridges were more rarely noted by observers. It

seems to have been fairly common for narrower ridges to be accompanied by furrows which were between one-third and one-half of the ridges' breadth. Arthur Young, for example, recorded examples of ridges six feet wide, which were separated in one instance by furrows two feet in width, and in another, two-and-half-feet.[35] Ridges also varied in height. The rather fragmentary evidence suggests that the centres of ridges might be anything from six inches to three feet higher than adjoining furrow bottoms.

Agriculturalists sometimes discussed the shape ridges should be in cross-section. Arthur Young believed that, in wet soils, ridges should be constructed so that in profile they would be smoothly rounded.[36] By contrast, however, some ridges made using spades were given firm, steep sides,[37] and in Erris, County Mayo, ridges were sometimes constructed to be asymmetrical in cross section. One side of the ridge was built up as much as 50 cms higher than the other. It was claimed that this helped drainage, and also that the sloping top of the ridge, which faced away from the prevailing wind direction, protected young plants.[38]

By 1800, the practice of making ridges in straight lines seems to have become widespread. Winding or 'curvilinear' ridges have been typically associated with medieval open-field systems, where the length of the plough and team meant that turning the oxen had to begin before the plough left the furrow, so creating a distinctive s-shape. S-shaped ridges were made in Ireland well into the nineteenth century. This was probably related to the use of the Long-Beamed Irish plough with its four horse team, which together made a very long assemblage. In 1802, Tighe claimed that curved ridges were still frequently made in County Kilkenny, and as late as 1844, Martin Doyle claimed that, 'The Common Irish, like the Norman farmer ... often [makes] a winding ridge'.[39] Doyle believed that the curves in these ridges were simply the result of the ploughman following any bends in the sides of fields, and that there was no good reason for the practice. Elsewhere, however, he noted that, 'In many of the clay counties in England, the old-fashioned form of a gently winding ridge and furrow is still preserved.' Here, he claimed, the intention was the same as that behind running drains obliquely across a slope 'to prevent a precipitate flow of the water.'[40]

Ploughs, like spades, were often used in making lazybeds (fig. 48), especially in hilly areas, or on poorly drained, shallow soils, which are especially common in the west of Ireland. One technique, recorded in several western counties,

48 Making ridges using at swing plough, at Slea Head, County Kerry, photo courtesy *An Clodhanna Teoranta.*

required great skill from the ploughman. The furrow slices turned to form each side of the ridge were made using the plough. This could be faster than using a spade for the same task, but it meant that each line had to be ploughed straight, without the guidance of a furrow slice turned previously. The whole ploughing operation required the same level of concentration as that required when a ploughman was making the first furrow turned (the score) in the middle of the area to be ploughed.

One recent local modification to wheel ploughs in County Leitrim was intended to help ploughmen with this task. Local blacksmiths fitted a board or 'block' on to the land side of the plough (fig. 49). This prevented the plough wobbling in the trench between ridges. The block pressed against the side of the ridge already turned on its left, and provided a guide for the ridge being turned on the right.[41] After the furrows forming the sides of the ridges had been turned, the mouldboard was removed and the plough was pulled along the bottom of the furrows created between the ridges. This broke up the soil at the bottom of the furrow, and this was then shovelled on to the top of ridge, both before and after the crop had been planted. It is as a clear example

49 Plough from Carrigallen, County Leitrim, showing the board fixed to
the side, used when making ridges.

of the persistence of ancient techniques into recent Irish farming practices, that
a few farmers in Counties Kerry, Cavan and Leitrim still use horse ploughs
(usually pulled by a tractor) to make ridges in areas of shallow, poorly
undrained soils (fig.50 a, b).

Ploughs and spades could be used together in a range of ridge making
techniques. One farmer interviewed in the 1980s in County Monaghan, John
Joe McIlroy, described how he made ridges on steeply sloping ground, using a
wheel plough and a spade. He only had one horse, and it was not strong
enough to plough uphill, so he made one half of each ridge using the horse,
ploughing down hill, and the other he dug, using a spade.

The spread of improved plough types, and the draining of land, led to a
standardization of ploughing techniques throughout many parts of Ireland
during the nineteenth century. Broad, very low ridges, known as 'flats' were

50 (a & b) Lazybed ridges constructed using a tractor and horse
drawn swing plough, Slea Head, County Kerry, 1970s.

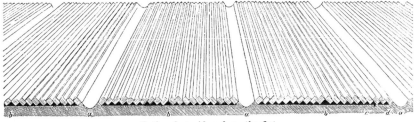

Gathered-up ridges from the flat.

a a a Three open furrows of two ridges.   *b b* Crown of two ridges.
  a b Furrow-slices lying from left to right.   *c d* Furrow-brow.
  b a Furrow-slices lying from right to left.   *d* Mould-furrow.

51 Ploughed 'flat': (a) hint ends; (b) crown of ridges, and line of 'scores', from *Stephens' book of the farm*, 3 vols (Edinburgh, 1908), vol. 1, p. 394.

ploughed. A flat was made up of about twenty furrow slices (fig. 51). The middle line of the flat to be ploughed was marked out or 'scored'. Furrows were ploughed up and down each side of the central score, the plough being turned at the end of each furrow, in a clockwise direction. A field was usually ploughed in a number of flats. Where two flats met, an open furrow was formed, which was narrowed and deepened by smaller furrow slices ploughed along each of its sides. This was known in English as a 'hint end'. This method of ploughing is still used at modern horse ploughing matches, where points are awarded for neat, straight furrows, as well as for the skill with which the central crown of the flat has been set up, and the hint ends. The same basic 'international' technique has also been used in tractor ploughing.

CHAPTER SEVEN

Harrows, Rollers and Techniques of Sowing Seed

H ARROWS AND ROLLERS can be used at several stages during crop cultivation. During the last two centuries, they have been most frequently
used just before, and just after, seed was sown. Harrows were used for breaking
up soil and weeding it before sowing, but also for mixing soil with newly sown
seed. However, other implements could be used for the same tasks. Some of
these, for example grubbers and horse hoes, were developed for use in 'improved'
systems of cultivation. On very small farms on the other hand, lumps of earth
were often broken up using wooden mallets known as mells, or in the south-
west of Ireland, heavy hand-hoes or mattocks, known as graffans (Irish:
grafáin) (fig. 52). Seed could be covered using a hand-rake fitted with iron teeth,
known as a *racán* (fig. 53), or simply by dragging a bush across the ground.
This last technique was so successful that some eighteenth-century agriculturalists recommended an 'improved bush harrow' made by twisting pieces of
white thorn around an old farm gate.[1] Most purpose-built harrows can be
easily distinguished from all these other implements by their construction.
Standard English terms can be applied to many of their parts (fig. 54).

Harrows have probably been used in Ireland for at least one thousand years,
but it is only in the eighteenth century that physical descriptions, and
illustrations, become detailed. Both 'improved' and 'common' harrows were
described, although the latter were often only discussed in terms of how they
might be improved. As late as 1830, harrows claimed to be in common use
among poorer Irish farmers were dismissed as 'villainous'.[2] In fact, as with
ploughs, harrows described as 'common' varied quite considerably from one
part of Ireland to another, differences even being noted within the same
county. In many areas, for example, observers distinguished 'light' and 'heavy'
harrows. In County Meath, in 1802, large harrows were used. These had outer
bulls which curved in towards a straight central bull (fig. 55a). In County
Kilkenny in the same year, however, the harrows described as 'common' had
straight bulls (fig. 55b). Both the Kilkenny and Meath implements have five
bulls. In County Tyrone, however, the harrows recorded as in general use in
1802 had only four bulls. The number of tines or 'pins' on common harrows

52 Mr Danny Cronin of Killarney, County Kerry, demonstrating the use of a grafán.

53 The children of Joseph Cunningham, Straleel, Carrick, County Donegal, holding a spade and a racán, in 1957 (photo by A.T. Lucas, courtesy of the National Museum of Ireland).

54 Wooden harrows, showing parts: (a) leaf; (b) bull; (c) tine or pin.

also varied. In Kilkenny the number cited was 25, in Meath 24, in Monaghan 21, and in Tyrone, 20.[3]

IMPROVED HARROWS

Experiments to improve harrows were well under way by the mid-eighteenth century, and aim of these was to produce stronger implements. In England, improvers warned that unless harrow frames were made strong enough, the

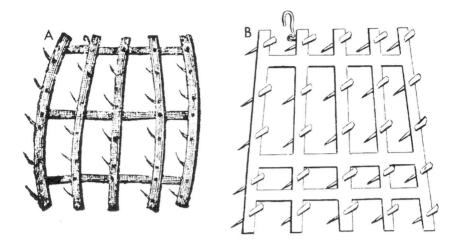

55 (a) Common harrow from County Meath, from R. Thompson, *Statistical survey of the county of Meath* (Dublin, 1802), p. 116; (b) common harrow from County Kilkenny taken from W. Tighe, *Statistical observations relative to the county of Kilkenny* (Dublin, 1802), p. 300.

implement could be broken apart, especially by the heavy work of breaking up ground. In 1769 John Wynn Baker produced harrows in his factory in County Kildare which were riveted on each side of the 'pin' holes, so that the wooden bulls would not split when the tines were inserted before use.[4]

A frequent criticism of 'common' Irish harrows was that when they were pulled through the ground, the rows of tines followed one another along a single track instead of each breaking up a separate piece of ground. Discussions about possible improvements also considered the best weight for harrows, the shape of the tines and the angle at which they should go into the ground. Tighe's discussion of the 'common' harrows of County Kilkenny raises most of these points:

> The tines being put in without system, often follow each other in the same track, leaving intervals of six or seven inches unstirred; their points incline generally too much towards the fore parts of the harrow, and their sides are turned to the front instead of their angles: by these circumstances the draft becomes difficult and the harrow is soon clogged with soil … Add to this that being made in one frame it is too heavy for the driver to lift up conveniently.[5]

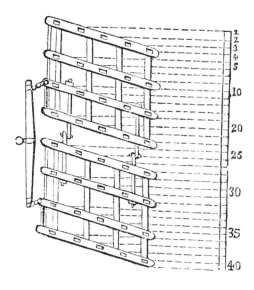

56 Rhomboidal, or Scotch harrows, from M. Doyle, *Hints to smallholders* (Dublin, 1830), p. 71.

The claim that on 'common' harrows rows of tines simply followed one another is questionable. Some details of their construction suggest a deliberate attempt to prevent this. Tines on the curved bulls of the early nineteenth-century Meath harrow, and on the Kilkenny harrow, where the bulls converged towards the front of the implement, would have been unlikely to follow exactly in one another's tracks. Another attempt to avoid tines following one another is suggested by the positioning of the hooks to which draught chains were attached on both the Meath and Kilkenny implements. The hooks were fitted to bulls left of centre of the harrow frames, and again made it unlikely that rows of tines would follow directly behind each other. This placing of the draught hooks would also have partially answered Tighe's criticism that tines were pulled through the ground flat side on. In these arrangements, tines that were square or rectangular in section would have met the soil edge on.

It is also doubtful whether any of the 'common' harrows described by early nineteenth-century writers would have been pulled smoothly enough through the ground for the tines to follow one another in firm, straight tracks. Even some modern horse-drawn harrows are linked to the harness trace chains loosely enough to allow the implements to bounce around as they are pulled, and some early 'improved' harrows were deliberately constructed to ensure that this happened. In both Counties Kilkenny and Clare, experiments produced implements which moved with a 'hustling latitudinal [motion] ... so desirable in rough ground'.[6] The County Kilkenny harrow was attached to a swingle tree by a loose ring to allow this sideways movement.

The concern that no two harrow tines should follow the same track can be related to the use of high narrow cultivation ridges. On these ridges, cross-harrowing would have been limited. Where harrowing was carried out in only one direction, however, there was a danger that the effects of the implement would be undone by the horses' feet compacting the earth as they walked up and down:

> Let the husbandman ... be cautious that he do not expect too much from [harrowing] ... or depend upon it in cases where he should have recourse to more powerful methods of breaking the ground. This is a caution the more necessary, because it is an error that present farmers very frequently run into ... When they have thrown in their seed they go over it with the harrow, and being sensible that the clods of earth must be broken, ... When they see once or twice harrowing does not effect it, they go over the ground again and again, till the feet of the horses have trod the soil in a hardness that is very unfit for the growth of anything.[7]

By the 1830s the 'rhomboidal' frames of 'Scotch' harrows were often recommended as the best shape to ensure that tines broke up the entire surface of the ground (fig. 56). Martin Doyle urged his readers to 'observe the tracks in which the pins move – all separately and distinctly, and at equal distances from each other, loosening every particle of the ridge over which ... [the harrow] moves'.[8]

The criticism of the County Kilkenny common harrows, that their tines were inclined too much towards the front of the implement, has to be balanced against the advice given by early nineteenth-century English writers, that tines on heavy harrows should have such an incline towards the front.[9] Harrows with tines arranged in a perpendicular position were thought to be less effective in breaking up clods of earth. A forward tilt allowed the tines to take a better hold on the ground as the harrows were pulled. Many early twentieth-century harrows had tines shaped, in the words of one County Antrim blacksmith, 'like a plough coulter'. A forward tilt was not advised for tines on lighter harrows, however, since this made it more likely that they would become clogged with weeds.

By 1767, John Wynn Baker's factory at Cellbridge, County Kildare, was manufacturing harrows made in several sections or 'leaves', hinged together. Baker pointed out that these harrows had an important advantage: 'The harrow, in general use in this Kingdom, is too often ineffectual in its operation, by its being made only in one frame, but mine being made in two frames, united together by what I call coupling bolts, they lie close to the ground, even in irregular places'.[10] This was especially important where plough ridges were narrow and pronounced. In County Galway in 1824, the double harrow was recommended as it 'lies better each side of a ridge'.[11]

The other main advantage claimed for hinged harrows was that each section could be easily lifted. This has remained an important consideration in harrow design, since even modern horse-drawn implements become clogged when in use, and must be lifted from the ground for cleaning. Many of these harrows have a handle attached to each leaf so that the operator can lift that section, while continuing to walk behind the harrows and drive the horses. The handles also allow pressure to be increased when the implement is crossing a high patch of ground. In this way soil can be collected and carried forward, and then shaken off in a hollow place. Most experiments in harrow design seem to have used hinged constructions, and later nineteenth-century evidence suggests an increasing use of hinged harrows throughout Ireland. (One retired farm worker interviewed in the 1980s, said that harrowing was much more tiring work than ploughing. A ploughman could lean on the handles of the plough for support, but this was not practicable with harrows.)[12]

In discussions on the improvement of harrows there was a lot of disagreement. For example, in 1802, some County Meath farmers argued that making harrows in several sections 'lessens the specific gravity; more than half the weight of the harrow being in any one point at the same time, and prevents its reducing the soil as quickly as it would, if the harrow was undivided'. Even the single-framed harrow used in County Kilkenny had sometimes to be loaded with stones to increase its weight.[13] This disagreement, like many others, seems to have arisen from a failure to recognise that different types of harrows are required for different tasks. John Wynn Baker's catalogue of 1767 advertised ten types of harrows, but although light and heavy harrows were distinguished very early, harrow types were probably not sufficiently separated by their two main functions: breaking up fallows (heavy work), and covering seed (light work). This was identified as a problem by at least one writer, in 1804: 'It is ridiculous to suppose, that any one [harrow] can answer all the various purposes, for which it is required, though how seldom do we find the farmer who has an idea, that one of a second form is requisite'.[14]

During the later nineteenth and early twentieth centuries most of the harrows used in Ireland were similar to those developed in England and Scotland. The different functions performed by harrows were reflected in increasingly specialized designs. Disc harrows, *zig-zag* harrows, spring harrows and drill harrows were widely used on even medium-sized farms by 1900. However, much less specialized implements also remained in use. Harrows with wooden tines, in use in medieval times, were still used in the nineteenth century. As late as 1949, Evans claimed that 'the time is still remembered when the old "six bull" wooden harrow was fitted with wooden tines or cows' horns'.[15] Bush harrows also stayed in use for a very long time. At least one late nineteenth-century text described the function of bush harrows as if these were still in common use. For both spreading manure and covering grass-seed, the light dense coverage of the ground surface by a flattened bush continued to be cheap and efficient, even after web or chain-link harrows were developed in the mid-nineteenth century. As with so many implements and techniques associated with 'common' Irish farming practice, we must be careful not to confuse cheapness and apparent simplicity with crude or inefficient husbandry.

DRAUGHT ANIMALS

From the earliest times horses were the most common draught animals used in harrowing. This is in contrast to ploughing, where oxen were widely used until the sixteenth century. It has been suggested that horses were preferred for harrowing because the draught required was not as heavy as for ploughing, and horses were faster than oxen.[16] The practice of attaching harrows to horses' tails is mentioned in several seventeenth-, eighteenth- and nineteenth-century accounts. However, the advice of some agriculturalists, that oxen should not be used in harrowing, was not always followed. One of the earliest clear illustrations of harrowing shows the implement being pulled by two oxen. There are also references in Irish literature to men and women pulling harrows. One of the earliest of these comes from the Annals of Ulster, where an entry for AD 1013 states that Gilla Mochonna, king of south Brega, 'yoked the foreigners under the plough and two foreigners pulling a harrow after them'. In very recent times, by contrast, the story of a 'butter-witch' from near

57 Wooden roller near Carnlough, County Antrim (WAG 1171 UFTM).

Comber in County Down shows at least a memory of harrows being pulled by people. The witch was shot while trying to escape from a vengeful farmer, but later tried to explain her wounds by claiming that she had fallen while 'trailing a harrow'.[17]

ROLLERS

Rollers have probably changed less than any of the other horse-drawn implements discussed in this book (fig. 57). As recently as the 1980s, many farmers still found horse-drawn rollers efficient, when modified so that they can be pulled behind a tractor. However, even these relatively uncomplicated implements provoked some debate amongst agriculturalists. Harrows and rollers were often used one after another. Rollers complemented the work of harrows, breaking ground into a fine tilth, and compressing seeds into the earth, but they were also used to make the surface smooth, so that care of growing crops, and harvesting, were made easier. They could be used for this last purpose, to a limited extent, even where land was made into ridges, although farmers were advised to roll across the ridges.

During the last three centuries, rollers could be made from wood, stone or metal. In 1757, a British farming text, republished by the Dublin Society, advised that only wooden rollers should be used, as those made of stone or metal would be too heavy, and compressed the soil too much. No diameter was given, but the length recommended was eight feet.[18] Most nineteenth-century writers however, claimed that wooden rollers were suitable only for light soil, although they were still valued for levelling potato or turnip fields after digging, and also for compressing earth around newly sown grain. Wooden rollers were still sometimes used in Western Ireland until the 1990s. Around the Bloody Foreland, in north-west County Donegal, for example, rollers were constructed from a solid block of hardwood, bound with iron straps at either end to prevent splitting. This was a construction recommended in 1893 by Thomas Baldwin, although he criticized the narrow diameter (*c.* one foot, or 30cms) of many wooden rollers, claiming that this made them difficult to pull, and liable to sink into hollows.[19] However, some other agriculturalists approved of smaller rollers, arguing that a larger diameter lessened the pressure exerted by the roller at any point. Many of the stone rollers (fig. 58) still found around the Irish countryside are narrow in diameter. Some surviving specimens are made of granite, which was the stone said to be most commonly used in the nineteenth century, although sandstone was also used. In 1844, Martin Doyle pronounced granite rollers to be 'very efficacious, as well as extremely cheap, the cost of one being about £1, with a scraper and frame, while the price of a metal roller ranges from £8 to £12'.[20] Stone rollers had the serious defect, however, that they were liable to fracture if the roller was pulled over uneven stone roads or yards.

The rollers most praised by agriculturalists were made from hollow cylinders of cast iron. As with wooden and stone rollers, these could be mounted on a simple frame, or they could be fitted with shafts, either for one or two horses. One late nineteenth-century text suggested that farmers should buy the iron cylinders from local foundries, and have them mounted on a frame by a blacksmith, and surviving specimens suggest that many farmers bought rollers with the dimensions recommended in 1839; five to six feet (1.5 m to 1.8m) long, and twenty to thirty inches (50cms to 75cms) in diameter.[21] Metal rollers had important advantages over those made from made

58 Bertie Hanna using a stone roller in the Ulster Folk and Transport Museum, *c.*1980.

from wood or stone. Their smoothness meant that clods of earth were less likely to cling to them, and even more important, they could easily be made up of two or more cylinders set side by side, which made them easier to turn, without churning up ground.

The weight of a roller could be varied by piling stones in a box set on the implement's wooden frame. Farmers also sometimes added their own weight, by standing or sitting on the frame as the roller was pulled. In 1893, one Irish agricultural writer enthused about a solution attempted by Amies and Barford of Peterborough. This firm produced rollers which could be filled with water, so allowing very fine adjustments to their weight.[22]

Rollers with spikes were probably in use in south-eastern England by 1700, and were being manufactured in Cellbridge, County Kildare, by 1767,[23] although a letter to the *Munster Farmer's Magazine* in May, 1811, which also included an illustration of a spiked roller, expressed the opinion that not many of the 'gentlemen' of County Cork had one, 'and it is very certain that no farmers are provided with it'.[24] The most famous of all spiked rollers, Crosskill's clod crusher (fig. 59), patented in 1841, was immediately recognised as very efficient. The cylinder was made up of a number of cast iron discs each 30 in. (75cms) in diameter, along the length of a 5 ft 6 in. (165cms) horizontal axle. Each disc was serrated around its circumference like a cog wheel. As the

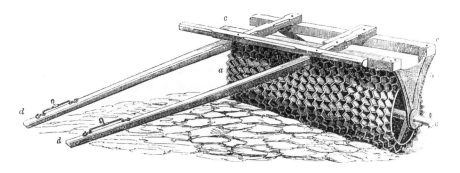

59 Crosskill's clod crusher, from Henry Stephens, *Book of the farm*, 2 vols
(Edinburgh, 1871), vol. 2, p. 21.

roller was hauled forward by horses, the discs revolved and large clods of earth
were broken down by the insertion of the teeth. Martin Doyle recommended
Crosskill's clod crusher to Irish farmers in 1844, and in 1863 the implement was
described as 'absolutely necessary on every farm'. In 1893 they were still being
used on large Irish farms.[25]

Like many other aspects of farm technology, however, even the slow
changes in roller design, and the low-key debates between agriculturalists
about the best shape and size for the implements were not relevant to many
small Irish farmers. In 1812, Wakefield found that throughout much of Ulster
the use of the roller was unknown, and in 1837, Binns claimed that throughout
the barony of Murrisk, County Mayo, there were no rollers used apart from
those owned by Lord Sligo.[26] Even where horses were used, many farmers
broke down soil with 'drags' made from pieces of wood nailed on to a wooden
frame. Arthur Young seems to have been referring to an implement of
this type, used at Furness, near Dublin, when he recorded a 'bull harrow …
that is … without teeth'. Horse drawn clod crushers were sometimes used well
into the twentieth century.[27]

SOWING

Grain seed could be planted in autumn or spring, but in the 1770s, Arthur
Young thought that in either case, many Irish farmers sowed their crops too
late.[28] There were differences in the time of sowing, however, and these
continued throughout the next century. In the later nineteenth century, for
example, turnips were said to be sown a fortnight and a month later in the

60 Sowing seed broadcast, County Antrim *c.*1906, Bigger collection
Ulster Museum.

west of Ireland than in the east.[29] During the period covered in this book, most small farmers sowed their grain and grass seed broadcast (fig. 60). Seed sown by hand was carried in boxes, baskets, skin trays or sowing sheets. Estyn Evans claimed that straw seed baskets were used only in south and east Ireland, where he thought they might be a survival from medieval baskets used in areas controlled by the Anglo-Normans.[30] The collection of the Ulster Folk and Transport Museum includes seed trays made by stretching a piece of cloth around a kidney-shaped wooden frame. However, simple pieces of sheet, and specially sown cloth bags, seem to have been the most common way for a sower to carry seed. The ability to sow seed in a wide even arc was prized, and a skilled man was often asked to neighbours' houses to sow for them. Some sowers held their sowing hand with the fingers slightly apart. The seed was thrown out through the spaces between the fingers, ensuring that it was more widely spread. Others flung handfuls of seed hard at the ground, so making it bounce and spread out over the ground.[31]

61 A seed fiddle in use (Ulster Folk and Transport Museum, WAG 3101).

In the early nineteenth century, machines were introduced which assisted Irish farmers to achieve more even broadcast sowing. One of these had a wide, narrow seed box with a lid, mounted across a light wheelbarrow frame. As the machine was pushed, a long axle fitted with brushes or iron teeth revolved inside the seed box and pushed seed out through evenly spaced holes. This basic form persisted over a long period and was used especially for sowing grass-seed. The length of the seed-box might be sixteen feet, and an even larger horse-drawn version was recommended to Irish farmers in 1839, but the implements do not seem to have become common.[32]

One broadcast sowing machine which did become very popular, however, was the seed fiddle (fig. 61). The early origins of these fiddles are obscure. It has been suggested that they were developed in England some time during the eighteenth century, but they are not described in either Irish or British standard farming texts before the early twentieth century. Fiddles are still sometimes used in Ireland, especially for sowing grass. Ensuring an even spread of seed is relatively easy. The sower takes regular strides, and pulls or pushes

62 A horse-drawn seed drill made by Ritchie of Ardee, County Louth,
Farmer's Gazette, 1847, vol. 6, p. 473.

the bow in time with each step. *Stephen's Book of the Farm* estimated that it was
possible to sow up to four acres of grain an hour using a fiddle, while three
could be sown easily.[33]

Experiments aimed at developing techniques of sowing, where seed could
be planted in evenly spaced rows have been carried out for at least three-and-
a-half centuries. The practice of 'dibbing' or 'dibbling' seed, which involved
making individual holes into which individual seeds could be set, seems to
have been first proposed in England by Edward Maxey in 1601, and became
fairly common in the southern English counties during the eighteenth and
nineteenth centuries.[34] Dibblers with rows of wooden points fixed to the
bottom of a wooden frame remained in use in vegetable gardening in Ireland
well into the twentieth century.[35] (The use of dibblers known as steeveens
(Irish: *stíbhíní)* for planting potatoes will be discussed in the chapter on that
crop.) The main significance of dibbing was that by planting in rows, hand-
hoeing amongst young plants was much easier, so that crops could be kept

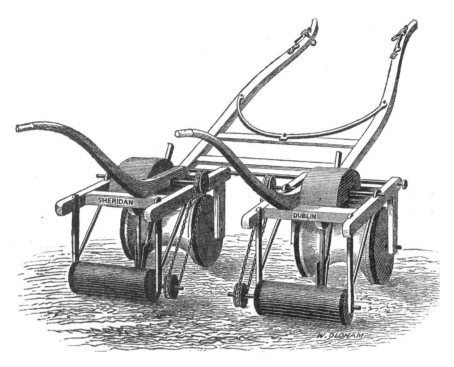

63 Turnip sower made by Sheridan of Dublin from
Farmer's Gazette, 1847, vol. 7, p. 187.

weed-free, and soil aerated. Dibbing therefore pointed the way to the major development in sowing seed during the last 250 years: the use of machines to sow seeds in rows or drills.

The most famous inventor of seed drills, perhaps indeed the most famous person in the eighteenth-century English 'Agricultural Revolution', was Jethro Tull. In fact, Tull was not the first to invent a grain drilling machine, but he certainly devised a practical one. Most horse-drawn grain sowing machines consisted of a large seed box fixed above the machine's wheels, with a series of tubes through which seeds fell at fixed intervals, so that they were sown in long straight, equidistant rows (fig. 62 and plate 4). This basic construction remained common, and by the 1860s seed drills were basically the same as those used in modern times. Small hand-operated seed drills for sowing grain were also developed, mainly for filling in spaces in drills where germination had not been successful, and drills were also made for sowing beans and other vegetables. From the evidence of surviving specimens, turnip sowing machines seem to have been fairly commonly used in Ireland in the later nineteenth and early twentieth centuries (fig. 63). John Sproule described the construction of a turnip sowing machine, in 1839:

> The machine drawn by a horse sows two drills at each turn … the seeds are put into cylindrical boxes of iron or tin, which are made to revolve during the time the machine is in motion, and being perforated in a line all round, the seeds fall through as the boxes revolve. Funnels are placed below each of these boxes, and through these the seeds are carried to the ground in tubes defended at their lower part by coulters, which make ruts in the ground into which the seeds fall. [The machine has] two light wooden rollers which follow each track of the coulters and cover the seeds … [and also a large wooden roller in front]. The effect of this roller is to flatten and compress these drills just before the seeds are sown.[36]

The spread of grain drilling machines seems to have been very slow throughout the eighteenth century. By the early nineteenth century, some machines were being manufactured in Ireland. A letter to the Cork Institute in 1813 records the purchase of a 'five row sowing and horse hoeing drill' from the Summerhill Manufactory, Dublin.[37] In 1839, however, drilling was said to have made little progress in Ireland, and this claim was repeated in 1854, but by the 1860s, machines for both broadcast sowing and drilling were said to be 'rapidly superseding sowing by hand'.[38] Even after this date, however, horse-drawn seed drills seem to have been confined to larger farms in established tillage areas.

There was widespread debate during the eighteenth and nineteenth centuries on the relative merits of broadcasting and drilling seed. In 1814, an article in the the *Munster Farmer's Magazine* listed the advantages of drilling grain seed:

1 Seed was saved.
2 Labour was saved. (If seed was ploughed in, a ploughman and a pair of horses could cover an acre a day. A seed drill with an operator and one horse could sow three acres.)
3 The increase of speed possible using a drilling machine meant that sowing could be carried out as soon as weather improved at all.
4 Crops resulting from drilling were claimed to be at least equally productive as those grown from broadcast seed.

5 Drilled grain ripened more evenly, because the seeds had been sown at an equal depth.

6 Drilled crops could be weeded much more easily than those sown broadcast.

The 'member of the Barrymore farming society' who wrote the article in which these advantages were listed, had discovered them after using a machine made by J. Barry of Glanmore, whom he recommended to his fellow farmers.[39] In 1839, however, John Sproule argued that while for the cultivation of green crops drilling was one of the most important improvements of modern agriculture, the value of drilling grain crops might have been exaggerated. He believed that drilling was advantageous in light soils but argued that, in the end, the value of a grain crop was determined by how much it tillered (or multiplied the number of its stalks) during growth. Drilling seed did little to encourage this proliferation.[40]

In 1844, Martin Doyle wrote that 'drill husbandry is the most perfect mode of culture, and we hope to see it widely extended in Ireland'. Even Doyle, however, allowed exceptions to this. Barley, he believed, did not require hoeing, which drill cultivation made easier, and probably tillered much better if the seeds were evenly spread over the surface by broadcasting, than if they were sown in drills where they were 'thrown so much together that they cannot tiller so freely'.[41] Another instance in which Doyle thought broadcast sowing might be best was when wheat was being planted in autumn, following a potato crop:

> In very stiff soils, and at the season of removing potatoes, the weather is often so moist, and the soil so adhesive as to render drilling difficult, if not impracticable; and then the most expeditious broadcast method must he resorted to. By this the seed is either ploughed under, in ridges on land in the fallow state (as is generally the case in the best wheat counties in Ireland) or harrowed in, after the ploughing, as is usual in England. The advantage in our judgement is greatly in favour of the Irish mode, which gives a far better and deeper covering to the seed, and thus prevents the plants from being thrown out in the spring, as frequently happens after severe frost.[42]

Doyle's discussion of this last case brings us back to points which will we hope to be emphasise throughout this book. In the end, techniques of sowing seed cannot be isolated from other cultivation techniques. The way in which ground was turned had direct relevance for the method of planting. High-crested work meant that seed tended to fall between furrow slices, and so grew up in rows. The same effect could be produced by the technique known as 'ribbing'. This was recorded by Tighe in County Kilkenny in 1802,[43] and was described by Thomas Baldwin in 1874 as a favourite mode of sowing wheat. 'The ground is made into ridglets or ribs, with a common plough stripped of its mouldboard, or with a ribbing machine. On cottier farms the ribs may be formed with the corner of a hoe or with a small shovel. The seed is scattered broadcast on the ribbed surface, and covered with a harrow'.[44]

'Common' techniques of sowing grain cannot simply be opposed to the drilling of seed as inefficient to efficient, or simple to complex. The spread of seed drilling machines was very slow even among 'improving' farmers. The reasons for this can be found in calculations of farmers, large and small, who took into account not only labour and draught requirements, and local environmental conditions, but also the ease with which new techniques could be fitted into their existing technical systems.

CHAPTER EIGHT

Potatoes

THE FAILURE OF the potato crop in the years 1845 and 1846 triggered the greatest catastrophe in recent Irish history, the Great Famine, and the economic, political and social results of the tragedy are still having deep long-term effects. Over one million people died of hunger and disease, and many more emigrated.[1] The Famine and its aftermath can only be understood by examining the wider political and economic structures of nineteenth-century Ireland. The rapid increase in Ireland's population before the mid-nineteenth century, the structure of rural society, and the development of industry or a lack of it, all made the position of the Irish poor very vulnerable, so that the failure of a single crop could bring devastation. While recognizing the fundamental importance of these issues, however, this chapter will concentrate on technical aspects of growing potatoes. In the end, these have implications for the big historical questions, but in this book we can only suggest possible connections, rather than trace them in detail.

Potatoes were probably introduced to Ireland from America in the late sixteenth century. Despite a tradition that Sir Walter Raleigh cultivated the plants in his garden at Youghal, County Cork, it seems equally likely that the first tubers may have come via Spain.[2] One of the earliest definite references to the crop dates from 1606, when Sir Hugh Montgomery's wife provided Scottish workers with land near Comber, County Down, on which to plant potatoes. The crop seems to have become a central part of the Irish diet very quickly. In 1689, John Stevens wrote that, 'None but the best sort … eat wheat or bread baked in an oven or ground in a mill; the meaner people content themselves with little bread, but instead thereof eat potatoes which with sour milk is the chief article of food'.[3]

Potatoes contain carbohydrates, protein, and mineral salts. Combined with milk, they can provide a liveable diet. Arthur Young recognised this in 1780, and also that apart from nutrition, potatoes had the advantage that large amounts could be produced from a small area. 'They have all a bellyful of food whatever it is, as they told me themselves; and their children eat potatoes all day long, even those of a year old'. 'One acre does rather more than support eight persons the year through, which is five persons to the English acre. To feed on wheat, those eight persons would require … two Irish acres, which at present imply two more for fallow, or four in all'.[4] It has been estimated that in modern Ireland an acre can yield up to nine tons of potatoes, enough to feed six adult men for a year. Before the Famine, consumption may have reached eight pounds a day per person.[5]

Potatoes were used not only for human consumption, but also as animal fodder, especially for pigs. Arthur Young viewed the eating habits of the Irish poor with affectionate derision:

> Mark the Irishman's potato bowl placed on the floor, the whole family upon their hams around it, devouring a quantity almost incredible, the beggar seating himself to it with a welcome, the pig taking his share as readily as the wife, the cock, hens, turkeys, geese, the cur, the cat, and perhaps the cow – all partaking of the same dish. No man can often have been a witness of it without being convinced of the plenty, and I will add, the cheerfulness that attends it.[6]

In the early nineteenth century, some observers believed that potatoes could provide a solution to the problem of providing food for Ireland's growing population. The large-scale movement of poor tenant farmers and cottars on to marginal land from the mid-eighteenth century was assisted by the relative ease with which potatoes could be grown. Just before 1845, there were about two million acres of potatoes cultivated in Ireland, most of these on plots of less than one acre.[7]

A great variety of potatoes have been cultivated during the last two centuries. Some of the most popular are listed below:

> *The Irish Apple.* This variety was grown by 1770. It was popular for its dry mealy consistency when cooked, and also because it kept well in storage, so providing food for the 'hungry months' of July and August, before the next year's crop had matured. However, it does not seem to have been grown much after 1846.

> *The Lumper.* One of the best known varieties of potatoes, which was valued for it's productivity, and also for its thick creamy consistency, which made it very filling to eat. However, it was very susceptible to blight, and now survives in very small numbers.

The Yam. This seems to have been a very popular variety amongst poorer farmers, as early as 1808. However, it was very susceptible to blight, and may have added to the scale of the destruction of the potato crop in the 1840s.

The White Kidney. A popular variety in the early nineteenth century, particularly because it could be harvested early, and so provided food during July and August.

The Cups. This was a luxury variety, grown as early as 1808. It was thought to be particularly nutritive, but was criticized because it 'stay[ed] too long in the stomach'.

The Champion. This variety was developed in 1863, and introduced to Ireland in 1876. Until 1894, almost 80% of the potatoes grown in Ireland were Champions, and they survived in use well into the twentieth century. A resistance to blight was one of the Champion's main advantages, but in 1890 it too succumbed to the disease.[8]

PLANTING POTATOES

It is very easy to get a potato plant to grow. (In the 1970s, one County Tyrone farmer remembered seeing the shoot of a young plant which had forced its way through the leather sole of a shoe thrown on a midden).[9] One way to obtain a new crop of potatoes was simply to leave some small tubers produced by the previous year's plants in the ground, so that they grew the following year. Arthur Young disapproved of this practice, which he recorded near Annsgrove, County Cork, and at Castle Lloyd, County Limerick: 'Potatoes they plant in a most slovenly manner, leaving the small ones in the ground of the first crop, in order to be seed for the second, by which means they are not sliced: sometimes a sharp frost catches them, and destroys all these roots'.[10] Planting in a more positive way could begin as early as February, and continue until May. Potato seed or 'sets' could consist of whole, or sliced up, tubers. There is also at least one recorded instance, where only the young shoots were used:

> I met in this town [Limerick] a certain reverend doctor, inventor of a new method of growing potatoes. This consists in cutting out, in spring, the shoots or eyes and planting them. It appears that the result is just as good as if the potatoes were cut up and planted, and with this benefit, that the tubers furnishing the shoots are still available as food – for pigs at any rate.[11]

64 James Cassidy and Paddy Cassidy demonstrating the use of a steeveen near Derrylinn, County Fermanagh in 1984.

Early nineteenth-century writers, however, generally accepted that large sets, or whole potatoes, led to increased yields.

Ridges, constructed using spades or ploughs, or both, were widely used for potato cultivation. Potatoes could be planted while ridges were being made. On lazy-beds, the tubers could be set on the untilled strips of grass which formed the core of these ridges, and sods then turned over on top of them. On the other hand, Arthur Young recommended the practice of setting potatoes in the furrows formed by lea ploughing, so that the next furrow slice turned covered the seed. 'In [the] … first ploughing, which should begin the latter end of February, or the beginning of March, the potatoes are to be planted.

Women are to lay the sets in every other furrow, at the distance of twelve inches from set to set close to the unploughed land, in order that the horses may tread the less on them. There should be women enough to plant one furrow in the time the ploughman is turning another.'[12] (The furrow slices in question were probably on ridges, but the practice of covering potato seed with a furrow slice could still be seen in the sandy *machaire* lands of the Outer Hebrides in the 1980s, where they were planted on flat ground.)

The spacing of potato seed varied widely. In 1807, Rawson suggested that potatoes should be set so that each plant would be surrounded by a square yard of soil. In 1814, in Dunaghy parish, county Antrim, however, sets were placed only six to eight inches apart, although farmers wanting to grow larger potatoes placed the seed from ten to twelve inches apart. At Athy, County Kildare, Arthur Young found that sets were placed fifteen inches apart, the seed being planted using a dibbler.[13] Potato dibblers varied in size from small hand-held sticks, to larger implements which were provided with a foot-rest. The best-known of these was the steveen (Irish: *stíbhín*) (fig. 64). Steveens were made of wood and had a foot-rest either cut into the shaft, or made from a separate piece of wood. They were used mostly in north Connacht and south-west Ulster. Even within these regions, however, potatoes could simply be set on the ground, and covered either with a turned sod, or by loose soil.[14] Farmers also sometimes used their spades to make holes in the ridge into which the seed was placed. This technique was known as kibbing. In 1814, Sampson recorded the practice in County Derry:

> In the operation of kibbing, the labourer is provided with an open bag of potatoes ready cut for seed, which he binds round his waist so as to have it ready to take out the seeds one by one. These he throws with great exactness into an hole, which he first makes by entering his spade forward into the soil, in a sloping direction; in raising the handle of the spade toward a perpendicular, a vacant space in the soil is left open, behind the spade-iron; into this vacuity the seed-potatoe is thrown … ; the spade being instantly withdrawn, the earth falls back to its level, and the potato plant is at the same time covered. An expert kibber will cover twenty seed in one minute.[15]

The late Mr Joe Kane, who planted potatoes in County Fermanagh by kibbing in the 1980s, said that it was particularly useful if the tubers had well-grown shoots since it ensured that these would be well covered, and so protected against frost damage.[16]

Just as potato seed could be planted at different stages in ridge construction, so manure could be added before, during, or after the ridge had been made. At Charleville, for example, Arthur Young found that the poor would 'sometimes put a little dung or [limestone] gravel on the grass, and plant it with potatoes'. In county Cork in 1813, on the other hand, it was claimed that 'the lower class' would sometimes dibble their potato seed into ridges, and spread the dung as a final stage. At whatever stage manure was added, however, contemporaries agreed that very large quantities were used. Arthur Young frequently noted that the poor used all their manure for their potato patches. Farmers sometimes rented land to cottiers already dunged, or let them have it free on condition that they manured it for potatoes.[17] The heavy use of farmyard manure persisted into the twentieth century. An agricultural inspector reported that in County Carlow, for example, 'the potato has the claim upon the manure heap and invariably gets a fair dressing of the best dung available'. At the same time another inspector criticized farmers in South Wexford for relying too much on farmyard manure, 'heavy dressings of which give unsound crops'.[18] The quantities of manure required when potatoes were grown in beds or ridges was a central criticism put forward by advocates of the drill cultivation of potatoes.

DRILL CULTURE

Drill culture was one of the most widely debated developments in eighteenth century farming. Drills are long, straight equidistant rows, and their regularity allowed systematic planting, care and management of growing crops, and harvesting. Potatoes were cultivated in 'raised' drills (plate 5). The ground was prepared, often by deep ploughing, cross ploughing, and harrowing. Raised drills could then be made using spades, lea ploughs and shovels, but 'drill' ploughs were also developed. These ploughs had a long narrow share or 'point', two straight mouldboards, and no coulter. When pulled through the ground, the mouldboards pushed loosened soil up in heaps on either side.

John Wynn Baker's factory at Cellbridge, County Kildare, was producing implements for drill husbandry in the 1760s.[19] At Lesly Hill, County Antrim, Arthur Young found that a Mr Hill had tried Baker's implements, but the experiment had not been a success. At Derry, County Tipperary, however,

65 Stages in making potato drills, as illustrated in a classic Scottish farming text: (a) ploughman making drills; (h, k, n, o, p) spreading manure from a cart; (r, s) planters; (u) ploughman covering drills, from Henry Stephens, *Book of the farm*, 2 vols (Edinburgh, 1871), vol. 1, p. 513.

Young found that potatoes were being grown in drills, apparently successfully, but without the standard equipment manufactured at Cellbridge. Stubble ground was first ploughed two or three times, and ploughs were then used to make the drill trenches. Dung and potato sets were laid down, and covered using either ploughs, or shovels. The drills were kept clean by constant moulding up with ploughs and shovels.[20]

An illustration taken from a nineteenth-century British text (fig. 65) shows the different stages in planting potatoes in drills. In 1893, Thomas Baldwin summarized these operations: 'In the hollows of the drilled ground, manure is spread; the "seed" or "sets" are placed over it at intervals of ten to twelve inches, according to the richness of the ground and the variety of potatoes grown; and both the manure and seed are covered by splitting the drills'.[21] In the early nineteenth century it seems to have been common to level the field with harrows after sowing. One reason given for this was that it made the plants grow straight up out of the ground, whereas if the earth was left heaped above them in drills, they would tend to come through the sides.[22]

66 William James McCann moulding up potatoes in the Ulster Folk
and Transport Museum (UFTM L1607/8).

The growing potato plants were weeded. Early writers complained that this
was not done enough, but some weeding by hand, and hand-hoeing of beds
and ridges were recorded. Several horse-drawn implements were developed in
the late eighteenth century for weeding drills. These included harrows, horse-
hoes, and grubbers. Most were constructed around a triangular frame whose
width could be adjusted so that the implement could be used on drills of
different dimensions, without damaging the growing plants (fig. 66). The
plants were also 'moulded up', sometimes several times. This involved heaping
soil up around the stems to encourage the growth of the tubers. Moulding up
usually followed weeding and grubbing, which loosened the soil between
drills. Shovels were often used for the task, but the double mouldboard drill
plough was the standard implement when horse power was available.

Agriculturalists, while admitting that lazy-beds had value on newly broken
ground, were adamant that potato drills were in general far superior. Indeed,

John Wynn Baker had to defend himself against the charge that the drill husbandry of grain and root crops was all that he was interested in:

> I have lately discovered, that some Gentlemen have formed an idea, that my undertakings have been principally calculated to establish the drill husbandry in Ireland. I shall beg leave to remove that idea, by reminding them, that the introduction of the drill husbandry is only one branch of the general plan.[23]

By the early nineteenth century, farming societies were energetically encouraging the cultivation of potatoes in drills. Potato drills were claimed to save labour, time, seed and manure. The crops were argued to be easily tended while growing, and more easily harvested. Tighe's dismissal of objections to drilling potatoes in County Kilkenny shows the determination with which the practice was promoted:

> The drill planting has made a slow progress in this district, though a general practice in Wexford; a few gentleman practice it, and some farmers, particularly in Galmoy, who learned it from the Palatines in Tipperary; some who have tried it, state ill-founded objections; they say saving manure is disadvantageous to the succeeding crop … they say the produce is not as great as in the old way, and the potatoes are too near the surface, and liable to be injured; these objections only spring from their defective mode of execution, by making the drills close, by not ploughing straight, and by not earthing sufficiently.[24]

Farming societies rewarded good examples of drilling. In 1810, for example, the Cork Institution offered a premium to 'The two farmers who shall cultivate in drills in the best manner, the greatest number of potatoes'. Small-scale projects were also encouraged. In 1814, the Institution gave a premium to a cottager, 'Elinor Leary of Ballyorban, for drill potatoes cultivated by herself and a boy, her son with a spade and shovel, the drills were marked with a line'.[25]

Members of the Cork Institution also questioned seventeen men who had won premiums for the drill cultivation of potatoes, on the techniques they used, and the merits of the system in general. Five of these men worked as ploughmen for 'gentlemen' and the other twelve were 'working farmers'. It was found that while several of the men had used double mould board ploughs to

67 Patrick Doran and his daughter making potato drills using a wooden Mourne drill plough, near Attical, County Down, in 1962 (UFTM L1864/5).

make the drills and cover the seed, others had used Scotch or Irish ploughs. The soil needed for moulding up had been loosened in a variety of ways with hoes, bush harrows, and in one case, using the local type of mattock, a graffan (Irish: *grafán*).[26] All of the men questioned agreed that drills left the ground in better condition than beds (or ridges) did, although in one case only one third of the manure was put in the drills that had previously been put in beds. One witness calculated that drills saved between one-third and one-half of the labour, one-third of the dung, and one-third of the seed which would have been used for cultivation in beds. The only instances where beds were thought better were in newly cultivated ground, or in very wet ground.[27]

It is difficult to discover the rate at which the drill cultivation of potatoes replaced the use of lazy-beds, but impressionistic evidence suggests that there was a large-scale adoption of the new methods during the early nineteenth century. In 1834, for example, it was claimed that, 'The drill has so completely superseded the lazy-bed that in the line extending from Dublin by Athlone to Galway, and thence by Ennis and Limerick to Dublin again, we saw but very few fields cultivated in the old way'.[28]

Nineteenth-century Irish foundries, and local blacksmiths, produced large numbers of metal drill ploughs, grubbers and horse-hoes, all for drilling potatoes. Wooden drill ploughs were also made, and in a very few areas, such as around the Bloody Foreland *(Cnoc Fola)* in north-west Donegal, some of these were still used in the 1980s. Around the south-eastern slopes of the Mourne Mountains in County Down, an evolution of drill ploughs occurred, which shows the ingenuity with which farmers would make use of older implements to take advantage of new techniques (fig. 67 and plate 6). 'Mourne' drill ploughs were probably developed from the Old Irish Long-beamed Plough, which was described earlier. Modifications to these ploughs probably began when the techniques of drill husbandry became known in the area. Rather than buy, or make, a double-boarded drill plough, local farmers adapted their lea ploughs for making drills by removing the coulter and lashing a piece of wood on to the plough's land side. This block acted as a second mouldboard, pushing loose earth up into drills. During the nineteenth century, the block, known as the 'false' or 'wee-reest', became more permanently attached to the plough, often lightly nailed in position and only removed when furrow slices were being turned. When metal swing ploughs became common in Mourne in the second half of the century, the ground-turning function of the local ploughs became obsolete. Coulters were no longer attached to the ploughs and false reests became permanent fixtures. This created an asymmetrical type of drill plough which continued to be made by carpenters and blacksmiths in Mourne until the 1950s.[29] Mourne type ploughs may not have been unique to that part of County Down. Similar asymmetrical drill ploughs were found in parts of north-west Donegal. We do not know if this was an independent development.

Despite the large-scale adoption of drill culture, the use of lazy-beds has also persisted in Ireland, especially on small farms situated in areas of marginal land. On many of these holdings family labour was more than sufficient to perform the work, and the saving in time claimed for drill cultivation was largely irrelevant, since as we have seen, ridges could be made by stages during the course of several weeks, or even months. Manure could be a problem if the family did not have a cow, but if dung was available from even one animal, there was sufficient to manure a crop. By using hand-labour, the farmer was spared the expense of keeping a horse, and could ensure yields which were at least as great as those produced by larger-scale techniques.

68 A potato digging fork illustrated in the *Dublin Penny Journal*, 1834, vol. 3, p. 283.

HARVESTING POTATOES

In Irish folklore, the date most quoted for the beginning of the potato harvest was Lammas day (Irish: *Lá Lúnasa*). This could be the last Sunday in July, or the first Sunday in August.[30] It was sometimes thought to be unlucky to dig the crop before this date, or to reflect bad management by the farming family. Digging the crop could be delayed much longer than this, however. At Kildare, Arthur Young found that cottagers were obliged to clear their potato land by the first of November, after which it was ploughed by the farmer and sown with winter wheat. At Annsgrove, County Cork, however, where there was no such pressure, potatoes were left in the ground until nearly Christmas, while at Nedeen he found that 'the climate in these parts of Kerry is so mild, that potatoes are left by the poor people in the ground the whole winter through'.[31]

Potatoes cultivated in beds were dug out using spades or forks. One such fork was advertized in the *Dublin Penny Journal* in 1834 (fig. 68). The maker, 'J.M.' of 27 Frederick Street, north Dublin, described the implement:

> The flattened portion of each prong is about five inches in length, by one in breadth, thinner at each edge than in the middle, and with spaces of one inch between each prong. They are made of scrap or Swedish iron, (occasionally I make them entirely of steel, in which case they are very light and handy, and wear much longer than when made of iron). They are made with a prong to go into the handle of about four feet in length, and are found much more convenient for digging potatoes, and also for pointing borders in a garden, than any spade.[32]

In large fields, ridges or drills were often opened using a plough. A Mr Kenney described the arrangement of his workers in a field in County Cork in 1810. The potatoes had been cultivated in beds four feet wide, separated by trenches two feet wide:

The ends of the beds, inconveniently situated for ploughing out, are previously dug with spades, and the stalks cleared away ... Twelve labourers with spades are provided, each attended by an active picker. The bed to be cleared is measured into six equal portions, one of which is allotted to each pair of labourers. A common plough, so set as to penetrate to the deepest potato, makes a cut along one side of the bed. The pickers of the first pair of labourers follow the plough quickly, and having collected such potatoes as have appeared, return to the labourers, who are by this time at work back to back, each beginning at an extremity of the portion of the bed allotted to them. As the plough proceeds, the other labourers and pickers go to work in the same manner. The plough returning cuts off a slice at the opposite side, which undergoes a similar course of treatment ... A plough thus attended will accomplish an acre and quarter in a day, even when the crop is abundant.[33]

Horatio Townsend commented that this arrangement was extremely well managed, and an improvement on that used for drills, which was similar, except that the labourers were not employed in pairs. The arrangement of two men and pickers following a plough was also recorded by Rawson in County Kildare in 1807, although in this case two boys with light forks were also employed, to 'shake out' the potatoes to the diggers.[34]

Lifting potatoes by hand was dirty and tiring. The labour involved has been well described by several Irish writers, from Patrick MacGill to John McGahern.[35] However, the even spacing of drills and the relative looseness of the soil meant that the mechanized harvesting of potatoes was made much easier. Fittings were developed in the nineteenth century that could be attached to the front of a plough. These pushed the potatoes out of a drill as it was split open during ploughing. The fittings could be similar to very large paring shares, but more often seem to have been arrangements of iron bars fanning out backwards from the front of the plough. As the plough cut through the drill, soil passed through the bars, but larger tubers were pushed out of the drill.

Experimental potato digging machines were developed in England and Scotland during the early nineteenth century, but it was an Irishman, J. Hanson of Doagh, County Antrim, who patented one of the most successful designs in 1855.[36] (fig. 69) Two horses pulled this machine, and as they did so a broad knife blade, or 'plough piece', cut under the drills. The

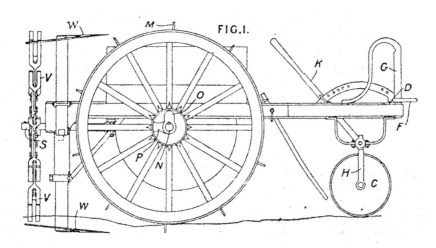

69 Potato digger, patented by J. Hanson of Doagh, County Antrim, in 1855.

earth and potatoes undercut by the blade were then scattered by revolving iron forks fitted to the back of the machine. The potatoes were usually prevented from being thrown too far by a net or screen fixed to one side of the forks. Hanson's machine became the prototype for many later designs. It meant that labourers had only to lift the potatoes left lying to one side of the drill. The speed with which the machine dug the crop, however, meant that a fairly large labour force was required for a fairly short time.

STORAGE OF POTATOES

The most common way to store potatoes during the nineteenth and early twentieth centuries was to make a potato pit (also known as a 'clamp', a 'bing', or in Irish, *poll-prátaí*) (fig. 70). The width of the pit varied between four and six feet, but it could be made to any length. The base of the pit was usually dug to a depth varying between several inches and one foot. The potatoes were put in to the pit in a gently rounded heap. Each layer of potatoes might be separated by a four to five-inch layer of dry earth, sand or ashes, and a shallow drain should be dug around the heap. *Purdon's Practical Farmer* suggested a more elaborate construction, with a row of drain tiles or pipes running lengthwise underneath the pit, and vertical ventilation pipes set into the heap at every three or four yards.[37] The heap of potatoes was covered with straw. Some farmers also used rushes or the stalks of the potato plants, but it was claimed that the stalks could injure the crop, since if not properly dried, they

70 Storing potatoes in a pit or 'bing', near Newcastle, County Down, *c*.1915 (UFTM, WAG 280).

would become mouldy and rotten.[38] The pit was completed with a covering of several inches of earth, beaten firm using the back of a shovel. When potatoes were needed during winter, one end of the pit could be opened and tubers removed. The pit was then resealed. Potato pits, if carefully made, were approved by agriculturalists. In 1814, for example, the Revd Mayne of Dunaghy Parish, County Antrim, wrote that, 'Since I have practised keeping my potatoes in pits in the field, I have never had what is called a curled potato. They are put up close and covered with dry mould only, and when taken out for use, are as well tasted as when newly turned out of the drill or ridge'.[39] In 1812, however, another cleric in County Cork complained that, 'The practice is lately become too prevalent of storing potatoes in numerous small pits in the potatoe-field, while the operation of digging or ploughing out continues. The farmer is too indolent afterwards to remove them speedily, and the ground is reserved for spring-corn. If this practice should continue, we shall soon see the country entirely divested of wheat-crops'.[40] Apart from this gloomy prediction, however, potato-pits were seen as the most efficient means to store potatoes. The alternative practice, of storing large heaps in a cellar or out-house, was actively discouraged after the 1840s, when it was recognized that these heaps were especially vulnerable to what became known simply as 'the disease'.

POTATO BLIGHT AND ITS TREATMENT

Potato blight *(Phytophthora infestans)* had appeared in America some years before it reached Europe. It was first reported in Ireland in Counties Wexford and Waterford in early September, 1845.[41] After a short time the threat it posed to the country became evident, as the Great Famine began. Most debate at the time, and since, centred on the famine relief measures that were slowly put in place, but investigations were also started to find the cause of the disease. Some of these were ludicrous, ranging from 'the introduction of guano manure to a curse from God incurred by Catholic emancipation and the government grant to Maynooth. Some blamed lightning for blasting the crop. Among the Connaught people it was believed ... that the northern fairies [had] blighted, or rather carried off the potatoes'.[42] Out of all this turmoil, however, two possible causes emerged as the most likely. These were that excessive wetness had rotted the crops, or that the decay was caused by a fungus. For several decades the first of these two possibilities gained most acceptance. The ventilation funnels in pits, advised by Purdon, were one preventive measure advocated in a government leaflet of which 70,000 copies were distributed.[43] Measures such as this, however, proved ineffective in preventing blight. By the 1860s, some agriculturalists were resigned to the idea that the disease was a permanent possibility. 'We need not enter into any speculations relative to the nature and causes of the potato "disease". Volumes have been written on the subject, and as yet, in many respects, we know as little about it as we did in 1845'.[44]

The blight was in fact a fungus, which attacked the leaves of the plant first, and then spread through the foliage. From this it could be washed down on to the tubers by rain. The use of copper sulphate, often mixed with lime, was well established as a mixture used to kill fungus in wheat and, in Europe, vines. In 1882, an accidental spraying of potatoes beside a French vineyard led to the realization that the mixture could also kill blight. It became known as the 'Bordeaux mix'. Later the 'Burgundy mix', made up of copper sulphate and washing soda, also became common. (Thomas O'Neill has shown that in fact an Irishman, Garrett Hugh Fitzgerald of Courtbrack and Corkaree, County

71 Experimental spraying of potatoes in Ulster during the 1930s (UFTM L 1483/7).

Limerick, had written to the Chief Secretary in Dublin Castle in 1846, informing him of the effects of copper sulphate on blight, but the letter was ignored.)[45]

In the last decade of the nineteenth century, a major drive to introduce the spraying of potatoes began, especially in the western counties which had been designated as 'congested districts' by the Westminster Parliament in 1891.[46] Spraying machines were developed, both to be carried on the back, or to be pulled by horses (fig. 71). By 1907, the Irish Department of Agriculture was offering farmers loans to buy horse-drawn spraying machines, and had instituted a scheme through which hand-operated machines could be hired. Through leaflets, placards and advertisements, both the Department and commercial firms publicized the benefits of spraying. The potato crop was sprayed first at the end of June, or early July, and it was recommended that a second spraying should be given three or four weeks later. In 1907, Department inspectors found that while spraying had only been practised to a limited extent in Leinster, in Connacht, Munster and Ulster it was widespread.[47] In the following years spraying became standard practice even on small farms (plates 7 and 8).

Potato cultivation remained of major importance in Ireland for some time after the Famine. Just before the Famine, about two million acres of potatoes were cultivated.[48] There was an initial decline in acreage after the late 1840s, but by 1859, 1,200,247 acres were cultivated.[49] This fell away during the later nineteenth century, and except on small farms in the west, by the 1890s potatoes had lost their dominant position in the Irish diet. The home consumption of potatoes, however, has remained much higher in Ireland than in Britain, and during the twentieth century, the commercial production of potatoes became highly organized. The cultivation of 'early' potatoes was encouraged by the Department of Agriculture. In 1905, it was reported that although 'There has been no rapid extension … the industry has assumed such proportions as to affect the English and Scotch markets, and this year the fields of growing crops were visited by large numbers of merchants from England and Scotland'. The infrequency of frosts in western Ireland assisted the development of the project. In Clonakilty, County Cork, in 1905, the harvesting of early crops was started on 3 June, and potatoes grown at Lisadell were sent for shipment to Sligo on 21 June.[50] During the second decade of the twentieth century, a successful seed potato industry was started in Ireland which continued to develop in both Northern Ireland and the Irish Free State. R.N. Salaman, who was responsible for developing around twenty varieties for the southern industry, provides a happy ending to the history of potato cultivation in Ireland. In the 1940s, he visited small farms in the west, where by this time the cultivation of seed potatoes of a 'well-organised, scientifically supervised and successful industry', and concluded, 'I have found it difficult to recognise in these knowledgeable, and intelligent growers, the direct descendants of those who, less than fifty years ago, had been degraded and exploited by the same potato'.[51]

CHAPTER NINE

Hay-making

HAY-MAKING BECAME WIDESPREAD in Ireland during the late nineteenth century. The belief that there was no need for hay in Ireland goes back at least as far as the eighth century when the Venerable Bede wrote that, 'Ireland is far more favoured than Britain by latitude, and by its mild and healthy climate. Snow rarely lies longer than three days, so that there is no need to store hay in summer for winter use or to build stables for beasts'.[1] Hay was certainly made on Anglo-Norman manorial and monastic estates, but it has been suggested that the small number of Irish words relating to hay-making shows the lack of importance of hay in Gaelic Ireland.[2] As late as 1838, one landowner claimed that in parts of west Connacht some people had not known that it was possible to dry grass to make hay.[3] However, hay-making does seem to have been common throughout much of Ireland by the later eighteenth century. Arthur Young found that at Gibbstown, Mr Gerard, 'one of the most considerable farmers in the country', had 100 acres of hay mown each year for his sheep and bullocks, while at the other end of the farming scale tenants with 15-acre farms around Mahon, County Armagh, would mow two of these for hay.[4] Cottars who did not have enough land to grow hay, often bought it. Young also noted the export of hay from Waterford to Norway.[5] The Dublin Society included hay-making as a topic for investigation in the county statistical surveys published in the early nineteenth century. Hay-making was reported in all the counties surveyed, and in County Dublin especially it was found that 'great hay farmers … abound'.[6] However, pasture and hay crops grew greatly in importance during the nineteenth century, especially when tillage began a long term decline after the 1830s. By 1904 hay was the principal crop in Ireland, after pasture.

Hay could be included in a rotation as an annual crop, or mown from permanent pasture. When grown as an annual, grass seed was often sown along with a grain crop. Grass was also sometimes sown after potatoes or along with flax. Clovers and Italian rye-grass were the most widely recommended hay plants. However, in early accounts, the most common sources of hay-seed seem to have been the sweepings of the lofts in farms or inns. This was condemned, not only because there was no attempt to ensure the quality of seed, but because it would contain a large proportion of white 'English' grasses, which were seen as little better than weeds.[7] Many Irish farmers, however, believed that the mixture of grasses found in permanent pastures produced the most nutritious hay. It was reported in 1802 that in County Meath farmers tried to ensure the widest possible variety of grasses by collecting portions of seed from neighbours and mixing these with some red and white clover seeds.[8]

Some observers noted that fine crops would grow on land which had been tilled but then simply left to throw up the natural herbage. Arthur Young thought this was probably due to the moist Irish climate: 'The amazing tendency of the soil to grass and yielding the succeeding summer a full crop of hay. These are such facts as we have not an idea of in England'.[9] In the first half of the twentieth century, most hay made in Ireland was mown from land which had been under grass for a considerable number of years. At the same time, however, the specialized production of hay-seed was well established. Grass seed production in counties including Antrim, Derry, Monaghan, Cavan and North Dublin meant that Ireland did not become a net importer of grass seed until the 1960s. In some cases, the production of grass seed for market, meant that livestock on the farm had to 'exist' on oat straw.[10]

MOWING HAY

Hay was mown in Ireland between July and September, or even October in upland areas and the west. Most agriculturalists were agreed that in Ireland mowing was started much too late in the year. Throughout the nineteenth and twentieth centuries there was a consensus that grasses were best cut just when they were coming into flower. After this, most nutritive matter went into the formation of seeds, while the stalks rapidly became more fibrous. Some loss was accepted as inevitable, where a mixture of grasses was grown, and each type came into flower at a slightly different time. Rye-grass, for example, bloomed earlier than clover, so that by the time the latter flowered, some of the nutrition in the rye-grass had been lost.

Arthur Young believed that the delay in cutting hay in Ireland was sometimes due to the late sowing of grass-seed, but in 1824 Hely Dutton alleged that the main reason County Galway farmers left grass uncut in fields they had rented for a year, in conacre, was a desire to get as much as possible from the land.[11] A similar motive was identified in a leaflet produced in 1904

72 Woods, Burgess and Key hay mowing machine, *Agricultural Review*, 1860, vol. 1, p. viii.

by the Irish Department of Agriculture: 'The delay in cutting obviously arises from the desire to secure an increase in the weight of the crop, but while the postponement of the cutting until an advanced stage of ripeness has been reached may ensure a slight increase in the yield, it also entails a serious depreciation in the feeding value of the fodder'. The same leaflet pointed out that not only would early cutting mean that what was lost in quantity would be gained in quality, but that crops would be less weedy, and – most important – there would be a better chance of good weather throughout the haymaking process.[12]

Throughout the nineteenth and early twentieth centuries scythes were widely used for harvesting hay, and the general history of scythes in Ireland will be discussed in the chapter dealing with the grain harvest. In the later nineteenth century, combined mowing and reaping were being manufactured by American and British firms. The most popular early machines seem to have been those manufactured by the firms of Woods, and Burgess and Key (fig. 72). The correspondence columns of the *Irish Farmer's Gazette* during the 1860s include some worried queries about the possibility of Woods machines becoming clogged, especially in bottom meadows where grass was soft and lush. Generally, however, the machines were welcomed enthusiastically. In July 1863, a Woods machine was demonstrated near Ballina, County Mayo:

Machines of that description have not hitherto been introduced into that
part of Connaught, and the recent trial was, therefore, quite a novelty in
this neighbourhood. [A] ... meadow of unusually heavy rye-grass and
clover was mown in splendid style, at the rate, we are informed, of about
an Irish acre per hour.[13]

One reason given for the spread of mowing machines during the 1860s was the
difficulty of obtaining scythesmen, In 1863, *Purdon's Practical Farmer* estimated
the comparative costs of mowing a field using a Burgess and Key machine, and
using scythes. It was assumed that the machine and its operator could be hired
for the work, while the farmer would provide the horses. The cost of this –
three shillings per Irish acre, or about one shilling and ten pence per statute
acre – was claimed to be much cheaper than hiring scythesmen, which was
estimated to cost seven shillings per Irish acre, or four shillings per statute acre.[14]

Irish farmers were encouraged by state institutions to buy mowing machines.
In 1904, the Department of Agriculture and Technical Instruction recommended the small one-horse mowing machines manufactured by Pierce of
Wexford, and the Wexford Engineering Company. Small farmers who did not
think that the size of their holdings warranted the outlay of capital required to
buy a mower were urged to combine with other small farmers to buy the
machines co-operatively.[15] Loans were also given to help with the purchase. In
July 1912, a scheme was started in western 'congested' counties to provide loans
for the purchase of a range of implements costing £6 and over. By September
1912, 110 mowers had been purchased under the scheme.[16]

HAY-MAKING

The main criticisms of Irish hay-making techniques, apart from late mowing,
were that the hay was turned too often, and that it was left in the fields too
long. Well made hay should be dried as quickly as possible, so that it retains
some green colour, and has a pleasant smell. By contrast, Arthur Young described
hay making or 'marring' near Lough Derg on the Shannon, where the hay
cocks formed little islands in the flooded fields, and the whole process of
drying might take up to two months.[17]

The first stage in hay-making everywhere in Ireland was to turn to 'shake
out' the new-mown hay. This was particularly important when scythes had
been used, as they left the cut grass lying in heavy swathes. Early mowing

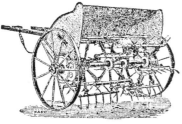 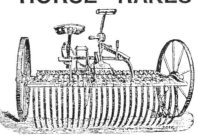

73 A hay tedder and horse rakes, *Farm and Home*, 11 July 1885, and 19 July 1886.

machines spread the grass more as they cut it. In 1844, it was recommended that four or five hay-makers, who could include 'women and boys', should follow each scythesman. The assumption that women would be part of a hay-making work team makes an isolated, rather sneering comment by Arthur Young that 'Irish ladies in the cabbins cannot be persuaded, on any consideration … to make hay, it not being the custom of the country'[18] more intriguing, but also more dubious. Shaking out the hay was often done with forks, although this was sometimes condemned as lazy, as it was more effective for the hay-makers to use their hands instead.

74 Paddy Cassidy of Lisnaskea, County Fermanagh, demonstrating how to make a hay lap in 1984.

Horse-drawn hay tedders, used for turning and spreading hay to dry, were developed in Britain in the early nineteenth century (fig. 73). A tedder designed by Robert Salmon of Woburn became the prototype for many later machines. By the 1860s, tedders were well-known in Ireland, where their introduction was considered a great boon to farmers. Some experts, however, thought the machines turned hay too roughly. Clover hay especially was thought to suffer, as it became brittle when drying, and the finer leaves were liable to be lost if the hay was shaken. But whether tedders were used or not, it was alleged that hay in Ireland was often turned too much. A description of hay-making practices in County Dublin in 1802 was particularly bleak: 'The hay is in general turned, and turned over and over again, until at length it is frequently caught by rain; then it must be dried, and this process of turning is renewed and continued, until it is devoid of either colour or smell'.[19]

Mid-nineteenth century texts advised that hay which had been cut and tedded in the morning should be raked into long 'wind-rows' in the afternoon. This could be done using hand-rakes, or with horse-drawn rakes, which were in use in England by 1831. The horse rakes which gained most common

75 Field of hay laps in the west of Ireland (C. Johnson (1901), p. 222).

acceptance were constructed on a frame set on two wheels (fig. 73). A row of curved teeth between the wheels collected hay until its own weight caused it to fall over. The slightly rounded teeth meant that the falling bundle of hay was made into a loose roll, which was praised because it allowed a free circulation of air through the mass. These rolls were sometimes likened to ladies' muffs, a description also applied to 'lap-cocks' which will be described below.

If the weather was very fine, the next stage in hay-making could be carried out in the evening of the day on which the hay had been cut, tedded and raked. This involved building the hay into small heaps known as 'field cocks'. In 1839, it was suggested that this work could be carried out by boys and girls under the supervision of the farmer or a 'confidential servant'. The young people were split into small groups which worked along sections of the field made up of either three or five ridges. Each group was divided into carriers and rakers. The carriers gathered the hay on to the central ridge, on which a row of cocks was to be built. A raker followed each carrier, collecting any hay left behind. It was suggested that where possible five people should work on each row of cocks; a carrier and a raker on each side of the row, and a person (preferably a more experienced worker) to build the cocks.[20] The technique of building the smallest field cocks shows a broad regional variation in Ireland

76 Charlie McAuley, Jim O'Neill and Malachy McSparran, making a hay stack at Drumnacur, County Antrim in 1963 (UFTM L3090/4).

between north and south. Throughout most of the southern half of the country the cocks were simply small 'bee-hive shaped' heaps of hay, but in the north (including counties Leitrim, Mayo, and Meath), lap-cocks (Irish: *gráinneog*, which is also a word for 'hedgehog') were often made (figs. 74 and 75). These were warmly approved of by agricultural writers. The technique of making a lap-cock was given in a nineteenth-century English household guide, where Irish lap-cocks were recommended to English farmers in wet seasons:

> The rakers follow each other, raking about four feet in breadth into the spread grass. A lapper, generally a woman, follows each rake, collecting a small quantity of hay evenly between her hands, which she places upon the raked round, doubling in one end as one presses it down, and with the right or left hand doubles in the other end, placing a small handful of loose hay across on the top.[21]

These rolls of grass, being open at the centre, did not pack together, and air could circulate and dry the hay evenly, and quickly. It was claimed that because of their construction, lap-cocks had been known to survive eight days of almost incessant rain without the hay being damaged.[22]

77 Moving a hay rick in the west of Ireland, photo courtesy of *An Clodhanna Teoranta*.

Lap-cocks and beehive-shaped cocks were usually spread out and remade several times over a period, which could last anything between two days and more than a week. Agriculturalists were united in agreeing that this was much too long. The next stage in hay-making involved combining two or more small cocks to make bigger ones (plate 9). In some areas this intermediate process of cocking or ricking was eliminated. Elsewhere, however, the larger cocks were spread out and remade until the hay was considered completely dry, so that this stage of the process could also last for a week or more. After this the cocks were again combined to make ricks or 'trampcocks' (fig. 76), because someone stood on the cock as it was being built, to compress the hay. Trampcocks varied in size, anything between 10 cwts. and a ton of hay being built into a single cock.[23]

By the beginning of the nineteenth century, some farmers had adopted methods which assisted the moving of larger amounts of hay collected in the later stages of hay-making. One practice, which seems to have been fairly common by 1800, was to put a rope around the back of a cock which was to be moved, and then to fix each end of the rope to a horse, either by means of a swing-tree, or directly to the horse's harness traces (fig. 77). This technique meant that two or more smaller cocks could be dragged along the ground at once. In 1802, one large farmer in County Meath extended this method so that

78 A Tumblin' Paddy in use in the 1950s, photo courtesy of the *Irish Times*.

even the largest ricks could be moved. He replaced the rope by a chain, and the draught was provided by six oxen.[24] Later in the century hay sweeps or collectors, developed in America and Scotland, became popular. Both American and Scottish sweeps were designed so that when full of hay they could be easily tipped over and the hay deposited in a heap. The sweeps which became known as 'Tumblin' Tams' in Scotland and 'Tumblin' Paddies' in Ireland were particularly common (fig. 78). These consisted of a row of six tines, each about three feet in length, spaced along a beam at intervals of about eighteen inches. Two handles joined to the beam allowed the farmer to control the sweep. A horse pulled the sweep with draught chains which were attached to swivel links at each end of the beam. The chains helped to hold the hay being collected, while the swivels facilitated unloading. This was achieved by allowing the points of the tines to catch in the ground, so making the sweep tip forward and turn completely over, releasing its load as it turned. When a full revolution had taken place, the farmer took hold of the handles again and began to collect another load. Sweeps could carry almost half a cart load of hay.

79 Ruck shifter on a farm at Gannoway, County Down, in the 1950s. The farmer, Mr William Lyons, is on the right, photo courtesy of Mrs Dolly McRoberts.

Irish hay-making practices were especially criticized because of the length of time taken between making trampcocks, and building more permanent stacks. Farmers were accused of developing a false sense of security once the largest field cocks had been. It was alleged that hay could be left in these ricks for weeks, or even months, by which time their bottoms, tops and sides were often completely rotted. In 1802, Hely Dutton claimed that the great hay farmers of County Dublin often sold this ruined hay to unsuspecting buyers at Dublin's Smithfield hay market:

> all the bad and refuse hay of the farm … is lapped up inside of each roll of hay, and loaded for Smithfield. The frauds that are practised in loading hay for this market call loudly for redress; it is a very common practice with many hay-farmers to shake a little fine hay on the ground; then a quantity of bad is shook evenly over this, and lapped up and loaded for market. In the market the farmer's man stands with a handful of hay, drawn from some part that has not been doctored.[25]

Hay which was kept for use on the farm could be stacked in a yard or haggard, or simply in the corner of a field. In the later nineteenth century, hay was often brought to the stacks on rick-shifters (fig. 79). These were low carts with a

80 A stack cover, 1860, *Agricultural Review*, vol. 1, p. cxlix.

large flat body which could be tilted backwards. Field cocks or ricks were then pulled up on to the cart, being drawn by a rope or chain tied around their backs in a way similar to that used earlier in the century to drag smaller cocks along the ground. The ricks were pulled on to the shifter manually, using a winching mechanism fixed to the front of the cart.

Hay stacks could be built on either a circular or an oval base. The stack could be raised from the ground simply by spreading a layer of straw or sticks on the ground, but even on small farms stone stack stands were also often constructed. The stack was made by spreading the hay regularly, but keeping the centre well raised. When the hay had been built up to a height of several feet, either with parallel sides or so that they bulged slightly outwards, the roof was begun. This was conical and steep. After the form of the stack was complete, loose hay was pulled off the sides to make the surface more smooth and compact. The completed stack was thatched with straw and tied down with straw or hay ropes. Nineteenth-century farming journals contain many advertisements for rick covers, which gave additional protection to the hay (fig. 80). By the 1840s cloths were said to be 'much in use at gentlemen's places, but amongst the farming class they are still a rarity. No farmer should be without one, as in one season he might save more by having it than twice the first cost. By having one of these cloths a farmer can bring in his hay and rick it whenever fit to be made into tramp-cocks in the meadow.'[26]

Stacks could be damaged by damp which led to heating, over-fermentation and eventual rotting of the hay. It was claimed that these processes could at least be hindered by mixing salt with the hay as the stack was being built. The amount most generally recommended was one stone of salt for every ton of hay, although *Purdon's Practical Farmer* suggested that up to two stones per ton could be used.[27] Salting was not universally advised, however. Martin Doyle argued that hay given to horses should not be salted, since this would increase their thirst and hence their intake of water to undesirable levels,[28] and in 1823, the North West of Ireland Society *Magazine* claimed that, far from preventing dampness, salting hindered drying, as it would actually attract moisture during wet weather.[29]

By the early twentieth century, hay barns were widely in use (fig. 24). Many of these were iron or wooden-framed structures, covered with sheets of corrugated iron. State institutions such as the Department of Agriculture and the Congested Districts Board gave financial assistance to farmers to build these barns. This policy was so successful that as one British text commented, hay barns were 'nowhere in this country so fully appreciated as in the Emerald Isle'.[30]

Many agriculturalists were insistent that Irish hay-making practices revealed a total lack of concern for the loss of nutritional value of the crop through exposure to rain, sun and wind. Rain especially was seen as harmful, because it washed nutrients out of the crop. In 1874, Thomas Baldwin calculated that losses during hay-making in Ireland amounted to 20% of the total value of the crop. Baldwin estimated that there were about 1,500,000 acres of meadow in Ireland, which produced about 2 tons per acre, or 3 million tons in all. Taking the current value of hay to be at least 50 shillings a ton, he calculated the total potential value of the hay crop to be £7,500,000. Of this, he claimed, £1,500,000 was lost by bad management.[31]

As with many Irish farming practices which were condemned by agriculturalists, the farmers themselves could defend their methods. Their main aim throughout the hay-making process was to prevent over-heating, and therefore over-fermentation of the crop. In 1808, it was reported that in County Clare, if a farmer put his hand into a cock and found it in the least warm, the hay would be immediately spread out again.[32] Agriculturalists argued that gentle heating of hay could in fact be beneficial, as the fermentation which this produced converted starch in the hay into sugar. In some areas in England, farmers allowed heating to continue until the hay had turned brown, and acquired a 'strong burnt-like smell'. Some Irish agriculturalists conceded that this was going too far. Hely Dutton thought it an example of 'English obstinacy'.[33] Over-fermentation meant that the sugar content of the hay could be converted first in to alcohol, and then into acetic acid.[34] Extreme over-heating could lead to spontaneous combustion in stacks. Irish farmers, defending their attempts to prevent hay from heat or fermenting at all, claimed that the ill effects of these processes were more likely to happen in Ireland than in England, since Irish grass was more luxuriant. In 1802, Richard Thompson recorded the case of a former bishop of Meath which was cited as evidence for this. The bishop 'pursued the English mode of hay-making, the consequence of which was, that about sixty ton of his hay took fire in Ardraccan haggard'.[35] An even more spectacular combustion was described in *The Farmer's Gazette* of 1847:

> We remember once seeing 500 tons of hay all but consumed by spontaneous combustion, on account of the new-fangled economy of a steward, who fancied he would conquer the elements by which nature surrounded him, nor did he fail without loss of life, for on lifting off the tarpaulin cloth by which his burning hay-rick was covered, two men were deprived of life.[36]

The *Gazette* used this report to make a general point.

> Very many who endeavour to effect reforms in our husbandry think they can do so by holding up the examples of our excellent friends [in Britain] … This, no doubt, is plausible and praiseworthy, provided it were done with due regard to the difference in the circumstances of soil, climate, etc., of the three countries; but these differences were never taken into account by the men brought into this country for its enlightenment, and although they did much good in introducing any change, for none could be for the worse, they have latterly proved a sad drag on improvement, as in point of prejudice against anything new or unknown to themselves, they are '*Ipsis Herbernis Hiberniores*'.[37]

CHAPTER TEN

Flax

Flax has had a very long history in Ireland. Flax was probably cultivated in the Bronze Age, and linen clothes and church vestments were commonly referred to in the literature of the Early Christian period. In the later eighteenth century, some flax production was part of the small farm subsistence economy. At Packenham in County Westmeath, for example, Arthur Young found that, 'The cottars all sow flax on bits of land, and dress and spin it, and it is woven in the county for their own use, besides selling some yarn', while at Hampton, County Dublin, they 'sow enough [flax] for their own use, not enough for manufacture'.[1] During the last three centuries, however, flax cultivation has been closely linked to the large-scale commercial production of linen. Given its economic importance, it is not surprising that, as John Sproule claimed in 1839, 'The cultivation of flax in this country has engaged a considerable proportion of the attention of the legislature as well as the farmer, and more laws have been framed relating to it than to any other branch of husbandry'.[2] In 1710, a Board of Trustees of the Linen and Hempen Manufactures was established, and some of the earliest premiums granted by the Dublin Society were given for flax, and flax-seed. Nineteenth-century bodies included the Flax Improvement Society, formed in 1841, and the Flax Extension Society, formed in 1867. At the end of the century the Irish co-operative movement encouraged the development of local flax societies, and in 1900 also organised a conference in Belfast for flax growers, scutchers and spinners.[3]

Despite all this institutional activity, however, flax production followed the pattern which can be traced for most crops in Ireland during the last 250 years: an initial period of expansion followed by a slow decline. By the end of the first decade of the nineteenth century about 70,000 acres of flax were grown annually, and this rose to almost 175,000 acres in the early 1850s.

The American Civil War meant that cotton imports to Ireland were almost completely cut off, and the linen industry boomed. However, this level of cultivation was exceptional, a brief reversal of a long term decline that had started in the late 1830s. By 1900, there were less than 50,000 acres of flax grown, and during the twentieth century, the crop almost disappeared, apart from two temporary revivals during the First and Second World Wars. Reasons suggested for the decline include competition from exports of cheap Russian flax, the cost of labour, and the reluctance of farmers to commit themselves to the heavy work of harvesting flax so near to the grain harvest. It has also been suggested that the use of poor seed had led to a succession of disappointing crops.[4]

Within this overall pattern, one of the most remarkable developments was the increasing concentration of the crop in the province of Ulster. More flax was grown in Ulster than in the other provinces as early as the seventeenth century, and by 1809 the difference in acreages had become very marked indeed:

Province	Acres (Irish) of flax grown	Province	Acres (Irish) of flax grown
Ulster	62,441	Leinster	4,107
Munster	3,716	Connacht	6,485[5]

The decline of flax growing outside Ulster was spectacular during the later nineteenth century. By 1890 there were only 886 acres of flax grown in Leinster, Munster and Connacht combined.[6]

National organizations made periodic attempts to extend flax cultivation outside Ulster, both by giving grants and sending instructors to areas which were thought to have most potential. A lot of this activity was centred on Munster, where markets were organized so that northern buyers could purchase locally grown flax. These failed, however, and seem to have been marked more by suspicion and disputes rather than by awakening enthusiasm. In 1891, for example, 'an unfortunate dispute between the local Scutch Mill owners as to the relative claims of Clonakilty, Dunmanway and Ballineen to have a flax market was the means of preventing northern buyers attending the southern market.'[7] Contemporaries and historians have explained the failure of flax production in southern Ireland by the lack of markets and local scutching mills. The concentration of industrialized linen production in the north east was also of central importance. One less serious explanation for the absence of the crop in many parts of Ireland was that the work of processing the crop, and especially the task of soaking or 'retting' was seen as demeaning. In 1774, Hyndman quoted, in pidgin Irish, a saying whose sense was derived from this attitude:

Neal fiss egum nach mhaphushe an lenn

(*Níl a fhios agam nach maothódh sé an lion*)
I do not know but he would water (or drown) flax.[8]

Within Ulster during the eighteenth century, linen production became concentrated in areas where there was a constant supply of strongly flowing water for bleaching work, such as the Lagan, Upper Bann, Roe and Faughan Valleys, on the Callan river near Armagh, and around Ballybay in County Monaghan.[9] Before the nineteenth-century industrialization of linen, production of the fabric had important effects on social as well as economic life in these areas. A dual economy of farming and of spinning and weaving linen developed, and gave the 'weaver-farmers' of linen-producing areas a relative economic independence. This led to a lifestyle which Arthur Young considered characterized by insolence and indolence:

> They [keep] … packs of hounds, every man one, and joining, they hunt hares: a pack of hounds is never heard, but all the weavers leave their looms, and away they go after them by hundreds. This much amazed me, but assured it was very common. They are in general apt to be licentious and disorderly; but they are reckoned to be rather oppressed by county cesses for roads, etc., which are not of general use … They are in general very bad farmers, being but of the second attention, and it has a bad effect on them, stiffening their fingers and hands so that they do not return to their work so well as they left it.[10]

As we have seen, the linen industry also tended to lead to a concentration of tiny farms. On small farms and cottiers' holdings the family labour supply was large relative to the acreage cultivated, and this was well suited to the labour intensive work of cultivating and processing flax.

PREPARATION OF LAND AND SOWING FLAX

Flax grows well on a range of soils, from rich loams to sandy and peaty soils. Arthur Young dismissed a claim, which he said was common in the north of Ireland, that very rich land was not good for the crop. However, this has been accepted in many written and oral testimonies.[11] Most agricultural writers agreed that flax could be included at almost any stage in a crop rotation, provided that it was not the first crop to be sown on lea ground, which might not only be too rich, but also might be infested with destructive leather-jacket grubs. The crop is so efficient at taking nutrients from the ground, it has been generally advised that flax should not be grown on the same piece of ground more than once in every five-to-ten years.

Flax should be sown on a seed-bed of fine, firm tilth. In the eighteenth century, it seems to have been common practice to plough the ground several times. In 1774, Hyndman claimed that the more often the land was ploughed, the better, but that it should be done at least three times.[12] Nineteenth-century writers often advised one deep ploughing, preferably in later autumn, but at least two months before the land was sown. The ground was finely broken up just before sowing, using a 'strong, coarse harrow', and then rolled. Large stones could also be pulled off the surface by the harrows, and placed either in the furrows between ridges, or at the edge of the field, where they could be collected later. In Ulster, women and children would often walk or crawl over the field collecting stones. Large clods of earth were broken up with a spade or a mell (a wooden mallet). Most writers also advised that the ground should be rolled before sowing, or that if a roller was not available, that a heavy harrow, with its tines turned upwards, should be dragged across it.[13] Flax should be sown between late March and late April, but later sowing could sometimes be best, because the crop was less likely to be damaged by bad weather.

Some eighteenth-century Irish farmers grew their own flax-seed. Arthur Young recorded that, near Armagh, flax intended for coarse linen was allowed to stand unpulled until the seed bolls ripened. Generally, however, both farmers and agriculturalists agreed that seed saved from Irish flax was not as good as that grown in a less moist climate. For finer linens, farmers preferred American seed for 'light and mountain' land, but Young found that at the time of his travels in the 1770s, this source had temporarily been cut off by the American War of Independence. On heavy, clay land farmers preferred Riga, Dutch or Flanders seed, claiming that it produced more flax of better quality.[14] In the later nineteenth century, the term Riga was applied to Russian seed, and this and Dutch seed were the two most common types grown. In 1904, Russia and Holland were claimed to provide almost all of the seed used for flax crops in Ireland.[15]

Throughout the nineteenth century, two-and-a-half to three bushels of seed were recommended per acre. Most writers agreed that flax was best sown broadcast. Drilling flax seed was usually only advised where the crop was being grown for seed rather than fibre. Drilled flax required less than half the quantity of seed used in the broadcast method, but it also meant that there tended to be a relatively wide space to the side of each plant, which encouraged the development of short side stems whose fibre was of very limited value. However, sowing flax-seed evenly using the broadcast method was even more difficult than sowing grain. Flax seed is small and light, and easily blown about by a breeze, so a calm day was recommended as the best for sowing. An added difficulty for the sower was that the shiny seed tended to slip too easily from the hand. Several writers advised that flax, like all plants with small seeds, should be sown thinly at first, and the remaining seed then used to thicken up the distribution. Seed barrows were sometimes used for sowing, but the most common mechanical aid was a type of seed-fiddle, which may have been the prototype of those already described in the chapter on sowing grain. The fiddle in question was the 'Little Wonder', patented by H. Adams of Garry, Ballymoney, County Antrim. An early twentieth-century description of this machine does not seem very different from those of grain-sowing fiddles.

> This machine, which is slung over the shoulder of the operator, consists of a hopper, from which the seed is delivered on to a disk, which is made to revolve by means of a string attached to a bow, hence the name 'fiddle'. The string is passed round a drum carrying the disk, and by moving the bow the disk is rotated and the seed thereby scattered. Whether sown by hand or machine, the land should be marked off in strips, each the width of a cast – 12ft. where the fiddle machine is used.[16]

A slightly different mechanism was used to revolve the disk on a sowing machine donated in the collection of the Ulster Folk and Transport Museum. On this machine, the disk was turned using a handle attached to one side. This sower is slightly smaller than the 'fiddle' type, and the donor said that it was used only for sowing flax.[17]

Harrowing and rolling the newly sown ground remained the most common method of covering seed. Light seed covering harrows were used, sometimes being pulled first along the field, and then across it. In the early nineteenth century, Mr Besnard, 'Inspector-General of the Linen and Hempen Manufacture for the Provinces of Leinster, Munster and Connaught', also recommended the use of 'a pretty large thorn bush, whereon you fix a weight of timber sufficient to make the thorns enter the ground'.[18]

81 Robert Berry and Robert MacDonald pulling flax in the
Ulster Folk and Transport Museum.

Care of the growing flax crop consisted mainly of weeding. If the ground
had been well prepared, it should be relatively weed-free, but the small leaves
on growing flax plants meant that other plants, including weeds, could grow
amongst the crop. Most writers advised that weeding should begin when the
flax plants were about four inches high. By the time they had grown to
between six and eight inches, the stalks were thought to have become too
brittle to regain an upright position after having been easily flattened by
weeders. During the eighteenth and early nineteenth centuries weeding was
usually done by women and children who crawled over the ground on their
hands and knees. Near Armagh, Arthur Young was told that ten women could
weed an acre of flax in a day.[19] Apart from minimizing the extent to which
plants would be flattened during weeding, the main concern was that pulling
out weeds would not disturb the roots of the flax.

PULLING AND RIPPLING FLAX

Flax grown for manufacturing linen was pulled by hand until the 1940s (fig. 81
and plate 10). Flax fibres in growing plants are very difficult to cut, and cutting
the stalks wasted the good-quality fibres near the roots. After pulling, the only

82 Rippling flax near Toome, County Antrim, *c.*1920 (UFTM, WAG 1024).

major processing operation carried out on the farm was steeping or 'retting' the flax. Flax grown for fibre was usually ready for pulling about one hundred days, or fourteen weeks, after sowing. The crop was pulled about three or four weeks after the first blossoms had appeared. Flax grown for fine linen was pulled earliest, but if pulled too soon the fibres were soft and weak, the yield was low, and the fibres might rot when retted. It was also generally agreed, however, that if left too long, the flax was only good for making coarse linen. Most writers advised that the plants should be pulled when the bottoms of their stems had begun to turn yellow and to lose the small leaves which grew up the entire length of the stalks. The development of the seeds inside the bolls at the top of the plant stalks could also be used as a test of readiness. It was agreed that the seed bolls should have begun to turn yellow, and that when cut open should have seeds which were 'no longer milky but firm, and somewhat brown in colour'.[20] It was also recommended, however, that flax which had been twisted by wind or rain, should be pulled as quickly as possible. On the other hand, dry, sunny weather could make the plants brittle, and here again early harvesting was advised. Careful harvesting also involved pulling long and short, and coarse and fine, plants separately. If this was not done, many of the fine, short fibres would be lost in the later processes of breaking and scutching.

Ridges or ploughed flats were used as guidelines to mark off patches to be pulled. Pulling a single handful of flax is relatively easy, but harvesting an entire crop was a backbreaking task. Each handful of flax was grasped about half way up the plant stems and pulled upwards and slightly to one side, and some skilled workers could use both hands for pulling. Four handfuls of flax made up a 'beet' or sheaf. The rate at which flax could be pulled obviously varied between workers, but one recent estimate suggests that one worker could harvest an acre in seven days.[21] In Ireland, beets were almost always bound using bands made from rushes. Bands could be made from flax, but this was considered wasteful as the fibres might be damaged. Beets were usually about eight inches in diameter, but careful farmers made them only four inches wide. This lessened the danger of heating inside the beet before it was steeped, and also meant that during steeping, the 'retting' process was more likely to be even.

It was generally agreed that a single flax crop could not produce both good seed and good fibre. Very little flax was grown for seed in Ireland, but the bolls were sometimes removed for animal fodder. The most common technique of removing seed was known as rippling (fig. 82). A 'rippling comb', made up of a set of iron teeth about 10 in. (25 cms) long was fixed to fixed across a small bench or plank. Handfuls of flax were pulled through the rippling comb. The seed bolls that were pulled off fell on to a sheet spread beneath.

Seed intended for fodder rather than planting could be rippled from the flax within twenty-four hours of pulling, but if it was intended for sowing, the flax was dried, by first being spread and then made into small sheaves and stocks. If the weather was fine, the seed could become dry and ripen after about a fortnight. When completely dry, the bolls were crushed by threshing or rolling, and the seed was then separated by winnowing. Some writers claimed that even where the seed was not to be used, rippling was essential to preserve the quality of the flax fibre, as the seed discoloured the flax and made it rot. However, early experiments by the Irish Department of Agriculture led their researchers to suggest that because of the difficulty of obtaining labour, and the uncertainty of the weather which made the flax harvest difficult

83 Steeping flax at the McSparran farm at Cloney, County Antrim, *c.*1915
(UFTM, WAG 1062).

anyway, rippling was not remunerative, and might damage the fibres. Some
Ulster farmers said that steeping the flax with the seed still on the plants was
good for the fibre, 'putting a shine on it.'[22]

RETTING FLAX

In the retting process the woody central parts of flax plain stalks, or 'shous',
were separated from the surrounding fibres. 'Dew' retting was sometimes
practised, where the flax spread out over a grass or stubble field, and
occasionally turned. Rain and dew provided the moisture which eventually
separated the woody stalks from the fibres. However, this system was more
common in England and mainland Europe than in Ireland. Most Irish farmers
retted their flax crops by steeping them in a dam or pond (fig. 83).

Flax which had been rippled and dried could be stacked, and retted during
the following spring, but most Irish flax was steeped very soon after pulling.

The temperature of the water used for retting was crucial, so steeping was usually started before mid-August, and continued only very rarely after mid-September. The quality of the water was also important. Hyndman advised farmers to use lake water, or water let into a river or stream. A river could also be used, he claimed, but water standing in a clay pit was very unsuitable.[23] Water draining from bogland was not recommended, as the water could stain the fibres, and water containing iron or lime was also rejected as unsuitable. Flax dams or ponds were usually dug near a stream or lake from which they could be easily filled. Dams tended to be long and narrow. One fairly typical dam in Ulster was about seventy feet in length and eleven feet in width. The narrowness allowed workers to fork the beets into the dam from both sides. It was not recommended that dams should be made deeper than four feet, as water below this depth was usually too cold at the bottom to allow easy retting.

The arrangement of beets in a dam varied. Early writers recommended that they should be set nearly upright. However, the beets were usually laid down almost flat, with the root ends only slightly lower than the tops. Two layers might be arranged in this way. Careful farmers kept beets of coarse and fine flax separate in the dam, as the coarser plants retted more quickly and should be removed sooner. The beets were weighed down with stones or sods, the stones being between 14 and 28 lbs in weight. The degree of compression was important. If weighted down too tightly, the plants retted unevenly. On the other hand, the retting process involved fermentation, and this often made the beets rise to the surface after several days, so that more stones had to be added. In cold weather some farmers added cow manure to the water to encourage fermentation. The cows themselves could present problems, however: 'Cattle love flax water – it's like beer … You have to keep them away, they go mad for it!'[24] Some farmers preferred not to let any water flow into the dam during retting, but it was also argued that a slow trickle might be advantageous, since it was generally accepted that retting in a large volume of water produced fibres of a better colour than was produced from flax steeped in small dams.

The length of time flax was kept in the dam varied between one and two weeks, depending on the type of flax, the quality of the water, and the temperature. Flax left too long in the dam became soft and tended to break during later processes in linen production. On the other hand, the woody stalks and fibres were difficult to separate if flax was under-retted. Some farmers believed that it was a worse mistake to under-ret than to over-ret. One test of the condition of retted flax was to break a few stalks in two places near the middle and about six inches apart, and take out a broken piece of shou. If it came away easily, the retting process was complete.

The beets could be removed a drag fork, or by workers actually getting in to the dam and throwing them out. Water in which flax has been steeped is very poisonous to fish, and also smells extremely bad. Even in the early nineteenth century, farmers who steeped flax in small rivers or lakes were condemned for destroying fish.[25] Farmers were advised either to let the dam empty itself by seepage, or to let the water into a stream during flooding.

Once beets had been removed from a dam, they were set upright in heaps to allow them to drain. After a few hours they were carted to a fairly bare field, where the flax was spread in rows to dry. The flax was turned several times during drying to prevent discolouring. Drying could be completed within six days in fine weather, but might take up to a fortnight. A farmer could tell that the plants had dried when a large proportion of the stalks formed a 'bow and string', caused by a contraction of the fibre from the woody stalk. After drying, the flax could be made into stacks until it was taken for bruising and scutching. However, although these latter processes were included as part of the dual farming-linen economy of many small Ulster farms in the late eighteenth, and early nineteenth centuries, they are related to textile production rather than crop cultivation.

Methods of flax cultivation changed less during the last three hundred years than those used in the cultivation of any other crop dealt with in this book. Until the Second World War, mechanization was minimal. Experimental flax pullers were developed in Europe during the 1920s, and during the 1940s flax-pulling machines were manufactured by James Mackie and Sons Ltd of Belfast, but by this time local flax cultivation had almost ceased. The most significant influences on Ulster farming during the last two hundred years did not come directly from flax cultivation, but from the related linen industry. Eighteenth-century commentators, in particular, condemned the effects of the dual economy of weaving and farming on the Ulster countryside:

> View the North of Ireland; you there behold a whole province peopled by weavers; it is they who cultivate, or they beggar the soil, as well as work the looms; agriculture is there in ruins; it is cut up by the root; extirpated; annihilated; the whole region is the disgrace of the kingdom; all the crops you see are contemptible; are nothing but filth and weeds. No other part of Ireland can exhibit the soil in such a state of poverty

and desolation ... the cause of all these evils ... [is] the [linen] fabric spreading over all the country, instead of being confined to towns.[26]

James McCully, writing a few years later, commented: 'To a stranger it must certainly appear paradoxical when he is told that the establishment of a prosperous manufacture, had almost ruined agriculture in the country where it is carried on'.[27] The paradoxical relationship between a flourishing linen industry and unsatisfactory farming practice can be paralleled in flax cultivation. Flax, closely associated with 'commercialization' and 'modernization', was one of the last crops to be mechanized, and was often grown on patches of land too small, or too steeply sloping, to be efficiently used in mechanized crop cultivation. More than any other crop, flax shows that no easy connection can be made between technological change and the advancement of farming.

CHAPTER ELEVEN

The Grain Harvest

DURING THE LAST three centuries, the main cereal crops grown in Ireland were oats, wheat and barley.[1] By 1700, exports of grain and flour to England, Scotland, and even France, were well established and this trade has continued, rising and falling with the relative yields of harvests in Ireland and abroad. There were periods when Ireland was a net importer of grain, but these were short term until the end of the nineteenth century. Grain cultivation was encouraged by the state. In 1758, for example, a bounty was granted on the inland carriage of grain and flour to the Dublin market, while Foster's Corn Law of 1784 gave bounties on grain exports. The period of the Napoleonic Wars, 1793–1815, was a boom time for Irish farmers, and although a severe slump followed the wars, exports of grain and flour doubled during the 1820s, and continued to rise steeply throughout the 1830s. The scale of exports during the 1820s led to the claim that Ireland had become 'the granary of Britain'.[2] Home consumption of grain was even greater than the amounts exported, and even during the 1820s, two-thirds of the grain grown was used in Ireland.[3] In the second half of the nineteenth century, however, there was an almost continuous decline in cereal cultivation. The table below shows this decline, but also shows the relative importance of each of the major grain crops in Ireland.[4]

Crops, 1850–1950 (hectares)

Year	Oats	Barley	Wheat
1850	857,000	130,000	245,000
1900	447,000	70,000	22,000
1950	388,000	66,000	149,000[5]

Oats were by far the most common Irish cereal. Oatmeal was a common element of diet, especially in the north. Oat straw was used for animal bedding, and until well into the twentieth century, it could also be used for fodder, as a substitute for hay. Oats were relatively easy to harvest. The light straw was easily cut, although this also made the standing crop vulnerable to high winds or heavy rain, which could flatten or twist the grain. Another problem, shared with wheat and barley, was that grain was lost if the crop was shaken during cutting. In Ireland this problem was made worse by the relatively late harvest. Compared to much of the rest of Europe, Irish summers are cool and moist, which means that grain ripens more slowly. In 1835, it was claimed that while oats and barley were usually cut in England during July, in Ireland they were not cut until August or even September.[6] The longer a cereal crop stands uncut, the looser the grain becomes in the seed heads. It was estimated by one nineteenth-century writer that the quantity of both wheat and oats lost in Ireland during the harvest by 'shelling and shedding' nearly equalled the amount of seed sown.[7] One unusual solution to the problem was suggested by a Belfast newspaper in 1750:

> As the corn harvest is approaching, it is recommended to the farmer, when high winds happen and he is afraid of his corn shaking, that he immediately break the stalks off all his ripe corn about the middle of the stalk; which may be done in a very short time with the hand: this will prevent the high winds doing much, if any, damage. None but the ripe corn will readily shake. This method has been pursued with good success in some parts of the country, but hath not hitherto generally prevail'd. A poor man having one ridge of rank corn, and no hands to cut it down, when a violent wind arose; he roll'd himself over that ridge, and then cut it down at leisure, and thereby preserved it from a great shake, which did considerable damage to his neighbour's grain. But had he only broke the stalks, as above directed, it would have been much easier afterwards to cut down.[8]

Despite the newspaper's approval, no evidence has been found that either technique suggested became widely used. However, one County Fermanagh farmer interviewed during fieldwork in the 1980s, described how he protected his tiny patch of oats from being shaken or flattened, by pushing sticks into the ground at regular intervals amongst the growing crop, and then loosely tying the surrounding grain heads to them. A more conventional solution to the problem, suggested by several agriculturalists, was that the grain should be cut three or four days before it was dead ripe. It was claimed, however, that even this advice was not followed by many Irish farmers.

During the period covered by this book, wheat was less well suited to Irish conditions than oats, preferring warmer, drier conditions, and growing best on

84 'Handsome Kate Kavanagh' from the Barony of Forth, County Wexford,
Irish Penny Journal, 6 February 1841, vol. 1.

heavy clay loams. In the late eighteenth century Arthur Young found that
some farmers in County Armagh believed that wheat would not grow there at
all. This was disproved by the Anglican primate who produced a 'very fine and
very clean' experimental crop.[9] Most wheat cultivation was, however, confined
to the sunnier, drier south-eastern counties. The grain was mostly used for
bread, but the straw was also valued. Wheat straw is harder and stronger than
oat straw. This limited its use for animal fodder, but meant that it was good
for thatching, and for the construction of a wide range of objects made in
Ireland from plaited or twisted straw. One nineteenth-century variety,
Whittington's, could grow to over six feet in height. Autumn-sown 'winter'
wheat was recommended as more suitable to Ireland than 'spring' wheat.
However, in order that the crop could survive heavy winter rains without
becoming waterlogged, it had to be cultivated on higher, narrower ridges than

other cereals. This, combined with the relative heaviness of the ripened grain heads, gave rise to special problems during harvest.

The cereal crop presenting most difficulties in harvesting however, was barley. Barley was used in brewing beer, ale and porter, and in the distillation of poteen (Irish: *póitín*) and whiskey. However, deciding on the right time to harvest the crop required careful consideration. If it was cut early, the grain suffered. The seed heads could shrivel, and since it had to be left longer in the field than other types of ripe grain, it was in more danger of overheating. If, on the other hand, the crop was left standing until it was fully ripened, the stalks became brittle and the heads were liable to break off during cutting. Over-heating remained a problem here also, since the crop tended to be weedy, and fermentation could occur if the grain was stacked, or even stooked, before the weeds had withered.[10]

Apart from oats, wheat, and barley, rye was the only other cereal cultivated in significant quantities in Ireland, mostly in sandy coastal districts. The grain was sometimes used for bread, for example in County Wexford,[11] and the straw was valued for thatching. In the late eighteenth century, rye had some importance as the first crop grown on reclaimed land by large-scale improvers. Rye presented relatively few problems in harvesting, and was the only major cereal which agriculturalists advised could be left standing until it was dead ripe. Like flax, rye was sometimes pulled rather than cut, especially if the straw was intended for thatching.

SICKLES AND HOOKS

Until the late nineteenth century, most grain was harvested by hand in Ireland, and sickles and reaping hooks were the most common harvest tools (fig. 84). Scythes were common by 1800, but until the late nineteenth century, they were mostly used for mowing hay. The main difference between Irish reaping hooks and sickles was that the former had smooth-edged blades, while the latter had blades with serrated, saw-like edges. By the seventeenth century the sickle had become one of the distinctive emblems of the Irish peasant,[12] and sickles continued to be the most common harvest tool until the nineteenth century, even after the earlier blacksmith-made tools were replaced by imported foundry-made implements. These later, mass-produced, sickles often had the teeth on their blades tipped with steel and the blade curved to form a semi-ellipse, elbowed just above the handle, this being generally accepted as

85 Paddy Storey from near Toome, County Antrim, demonstrating how to hold a sickle when reaping (UFTM, WAG 298).

the most efficient cutting shape. By the end of the nineteenth century, foundry-made sickles were used throughout Ireland. One specimen in the National Museum in Dublin (Reg. No. 1967.93), for example, was used on Inishbofin Island, County Galway. Agricultural implement makers, like Tyzack's of Sheffield, produced many patterns of sickles and reaping hooks. Two sickles advertised by the firm in the 1923–24 catalogue are described as 'Irish.'

The main advantage sickles had over smooth-bladed reaping hooks was that they did not require frequent sharpening. Hooks could cut grain faster than sickles, however. Some evidence from Scotland suggests that, although sickles were the tool most favoured by Irish reapers, hooks may also have been fairly common. It has been claimed that since the eighteenth century, Irish migrant harvest workers in Scotland favoured smooth-bladed hooks,[13] and in 1831 the agriculturalist J.C. Loudon claimed that reaping hooks had actually been introduced into western and south-western Scotland by Irish harvesters.[14]

Several techniques can be used in reaping grain with sickles or hooks. A method common to both implements involves taking a handful of straw in one hand and cutting it with the blade held in the other (fig. 85). Reaping using this technique can be either 'high', when only the ears and a short part

86 John Canning mowing oats with a scythe at the Ulster American Folk Park.
Liam Corry, on the right is poling the crop, holding it steady for the mower,
photo courtesy of the Ulster American Folk Park.

of the straw are cut, or 'low', when the straw is cut as near to the ground as
possible. These were the neatest ways of cutting grain, since the reaper, by
taking hold of the stalks just below the seed heads, could leave behind many
of the weeds growing lower down. Another faster method of reaping was possible
with smooth-bladed hooks, however. This involved hacking or slashing
standing grain, which might be held in position with a small fork made from
furze. In England this technique was known as 'bagging' or fagging', and large
'bagging hooks' were used. These larger hooks were also known in south-
eastern Ireland, but in general Irish reaping hooks were smaller.

In a large field, reapers were followed by binders who tied sheaves, each
made up of about three handfuls of grain. In Ireland, as in Scotland, reapers
generally worked along the ridges on which the grain was grown. A common
arrangement of reapers and binders in a large field was described by Martin
Doyle:

> The mode of reaping differs in Great Britain and Ireland. The British
> labourer first cuts a small handful for the band, which he deposits and
> then (cutting across the ridge) he lays across the band as many handfuls
> as will constitute a sheaf, which he himself generally binds. The Irish
> reaper, being a more social animal, has three or four companions on the
> ridge, according to its breadth; one cutting a passage along the furrow,

and ledging the several handfuls across it, and the others reaping a breadth of about eighteen inches each, every pair of reapers ledging together, but none of them preparing bands for their fair followers … In Scotland binders are generally provided, which is a great saving of time for the reaper, but in a much less proportion than in Ireland, where one binder (female) is allowed for two men.[15]

In fact, other accounts suggest that one binder had sometimes to work for more than two reapers, but Doyle's description makes the division of tasks between reapers and binders clear.

The arrangement of reapers across a field was also described in the preface to a late nineteenth-century Ulster poem:

The leader of the 'boon', or band is 'stubble-hook', so called from his being employed on the open plot next to those which have been shorn; while 'cornland' occupies the ridge next to the standing grain, and may be looked upon as the driver. The shrewd farmer generally chooses two of his best shearers for these situations. He knows that each reaper from the leader to the driver is supposed to keep about the 'making' of a sheaf in the rear of the hook immediately preceding him: and that, therefore on the 'stubble-hook' and 'corn-land' depend, in great measure, the amount of labour to be accomplished by the hooks at work between them.[16]

It has been argued, on the basis of Scottish evidence, that women workers were valued for reaping with sickles, their bodies being well adapted physically to the stooped position required of the reaper.[17] Irish evidence tends to follow a broader pattern identified throughout Europe since the medieval period. Both men and women reaped with sickles and hooks, but scythes were almost always used by men.

SCYTHES

The term used for 'scythe' in older Irish language texts, *spel*, may be a loan word taken from Middle English (fig. 86).[18] Scythes were used on Anglo-Norman manorial estates in Ireland. Labour services required of a *betagh* included the mowing of hay with a scythe. By the eighteenth century, references to use of scythes in hay-making had become much more common. The widespread use of scythes is indirectly suggested by Arthur Young's

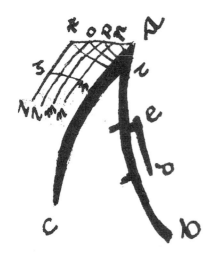

87 Cradle-scythe, drawn by Amhlaoibh
Ó Súilleabháin in 1831: (a, b) *crann na
speile*; (d, e) *na dornóga*; (a, c) *an speal*;
(a, f, I, n) *an cliabhán*; (fi, ol, pm, rn)
plata an cliabháin i gcuspa don speal;
from Amhlaoibh Ó Súilleabháin,
Cinnlae Amhlaoibh Úi Shúilleabháin,
vol. 3 (Dublin, 1930, p. 66).

erroneous explanation of the meaning of the term 'spalpeen' (Irish: sp*ailpín)*,
which was applied to temporary migrant labourers. Young claimed that the
term was derived from two words: 'Spal in Irish is a scythe a peen a penny, that
is a mower for a penny a day'.[19] In fact the term is probably derived from the
Irish *spailp,* meaning a spell or period, thus referring to the temporary nature
of the spalpeen's employment. Young's mistake, however, does suggest that
either he, or his informant, regarded the use of scythes as a typical task carried
out by these workers. Import records show that by the late eighteenth century,
large numbers of scythes were being brought into Ireland from Britain and by
1800, scythes were sometimes used for mowing grain as well as hay. In County
Kilkenny, Tighe found that 'a few farmers have occasionally mown their spring
corn: mowers do not like to undertake it, and say it spoils their scythes: it is
constantly practised by Mr Robert St. George, who advises that any kind of
crop, except wheat should be mown with a cradle scythe'.[20]

A story originating in the late nineteenth century and recorded in the
Mourne Mountains, County Down, by a local farmer, Joey Murphy,
emphasizes the relatively late use of scythes for harvesting grain. A migrant
labourer who had been working at the harvest in Britain brought a scythe back
with him. He demonstrated the speed with which grain could be cut using the
implement to a group of increasingly impressed neighbours.[21] This story
suggests that many small Irish farmers may have learned the technique of
harvesting grain using scythes while working in England and Scotland. (It also
has been claimed, conversely, that many Irish harvesters in nineteenth-century

Britain would not, or could not, use scythes, and this slowed the introduction of scythes as grain cutting implements in some areas.)[22]

Scythes with a 'cradle' attachment behind the blade to catch the newly cut swathe seem to have been first used in south-eastern Ireland. The diarist, Amhlaoibh Ó Súileabháin, made a diagram of a cradle scythe (Irish: *cliabhspeal*) in 1831 (fig. 87). His reaction to the implement, which he saw in use near Kilkenny, suggests the limited extent to which cradles were used:

> I saw a new arrangement on a scythe today, cutting oats for George Clinton. There was a cradle or cage mounted on the scythe … This scythe brings the [whole] produce of the ground with it, both straw and grain, for it cuts low, and lays it straight in the swathe … A mower with a cage scythe would cut as much oats or barley with this apparatus as four sickle men.[23]

Later in the nineteenth century the use of cradle attachments became more widespread. One specimen in the National Museum of Ireland (Reg. No. 1932:130), for example, came from Tuam, County Galway. An Irish National school-book, published in 1868, suggested that farmers could make their own cradles: 'A skilful man can mow corn very well with a common scythe … a few pieces of strong wire should be fixed to the sned [shaft] to catch the crop as it falls on to the scythe, and assist in getting it into a swathe'.[24] Other simple cradles were made by bending a twig to form a bow behind the scythe blade, or by stretching sacking or canvas over a curved piece of wire.

The most common scythe used in Ireland before the twentieth century had a long straight shaft or sned. Two handles were fixed to the shaft at right angles to one another, and the distance between these was set to suit individual mowers. The angle of the scythe blade to the sned was also adjusted for different crops, using a piece of wire sometimes known as the 'grass-nail', which linked the blade and sned. Improving landlords, returning migrant labourers and agricultural implement salesmen were probably responsible for the introduction of the wide variety of scythe types used in Ireland in more recent times. Scottish Y-shaped shafts of wood and metal, and English and American S-shaped shafts were widely known.

Dineen's *Irish-English dictionary* makes an interesting distinction between the *speal Ghaedhealach* (Gaelic scythe) and the *speal Ghallda* (modern or 'foreign'scythe).[25] The Gaelic scythe is described as having a rivetless blade.

This suggests that the distinction was between local blacksmith-made scythe blades and factory-made implements. However, some of the latter were also rivetless. The 1921 catalogue of Tyzack's Ltd of Sheffield illustrates the latter type, describing them as 'solid-back' scythes.[26]

Scythes require frequent sharpening when in use. Modern scythesmen recommend that this should be done every ten minutes. Both sharpening boards, or 'strickles,' and scythe stones were widely used in Ireland. Mowing with a scythe is a skilled task. Mowers generally cut along the edge of the crop, some swinging the blade towards the main body of standing grain, others away from it. As it approaches the grain, the 'heel' (the joint where blade and shaft meet) should be travelling along the ground. Mowing is generally agreed to be much quicker than reaping with hooks and sickles. English evidence suggests that a scythesman could cut grain four to five times faster than someone using a sickle, and more than three times quicker than someone using a reaping hook. More binders were required to follow a scythesman, to take full advantage of this increased speed of cutting, but overall only half the labour was required to harvest an acre of grain using scythes than that required when sickles or hooks were used.[27]

The late introduction of the scythe to the Irish grain harvest can be explained by the limited extent to which it could be used in common systems of cereal cultivation. These systems were developed within the very limited resources available to most Irish farmers, and also within the limits imposed by the mild, damp Irish climate. The high, narrow cultivation ridges that were widely used in Ireland well into the nineteenth century were a major obstacle to the use of the scythe, which required ground that was 'free from hills and hollows'. We have already seen that by 1800 some 'gentlemen' farmers in County Kilkenny were experimenting with sowing wheat on almost flat land, but this was not judged to be successful.[28] It was not until after 1850, when large-scale drainage schemes became widespread and cultivation ridges began to be flattened, that scythes could be effectively used.

The dominance of broadcast sowing techniques in Ireland also lessened the initial attractiveness of scythes. Grain crops sown in long straight rows by seed drilling machines could be weeded using hoes, but this was extremely difficult when seed was simply scattered on the ground. Weedier crops were best reaped with sickles or hooks, as harvesters could select the grain for cutting, leaving many the weeds behind. Scythesmen could not reap selectively, and this meant that farmers with sufficient labour could expect cleaner crops if they were cut using the smaller implements.

The lateness of the Irish harvest also hindered the introduction of scythes. The longer grain stands uncut, the more likely it is to be flattened or twisted by rain. Where this happened it was extremely difficult to use a scythe. It was also pointed out that grain left standing in a cool, moist climate, developed thicker straw. Irish barley straw, it was claimed, was too thick to be cut using a scythe. On mainland Europe, lighter, spring-sown grain, particularly oats, were the crops on which scythes were earliest used.[29]

Some Irish farmers claimed that mowing grain with a scythe meant that more seed was knocked out of the grain heads. In 1863, a correspondent in the *Irish Farmer's Gazette* claimed that in a test of scythes against 'fagging hooks' on an area of twenty acres of wheat, the hooks had proved cheaper, and had knocked less than half the wheat seed lost by scything.[30] Similar claims were commonly recorded in eighteenth-century England and Scotland. Some observers, however, argued that the reverse was the case, and that scythes actually shook grain less than sickles or hooks. Martin Doyle believed that the extent to which grain was knocked out depended more on the skill of the harvester than on the tool used.[31]

Differences between agriculturalists may have been based on observation of different techniques of reaping with sickles or hooks. Where the slashing or 'fagging' technique of reaping with a hook was employed, the grain could be badly shaken. The technique where grain stalks were held in one hand while being cut with the other would almost certainly have shaken the grain much less. In those parts of Ireland where this technique was employed, less grain would have been lost than if the crop had been cut with a scythe.

It seems that scythes only became attractive to most Irish farmers as grain-harvesting implements when drainage schemes allowed the old cultivation ridges to be flattened, systematic weed control was developed, and harvesting was begun earlier. The main advantage of using scythes – an increase in cutting speed – only became significant when farming was large scale. Many farmers on smaller holdings, where labour could be adequately provided by family members, neighbours, and some hired help, preferred to continue using the neater harvesting methods possible with sickles or hooks. This was a clear example of an apparent improvement in technology which was not always the best method for the Irish small farmer. And even as conditions were changing to allow the efficient use of scythes, these were already being technologically superseded by the development of reaping machines.

REAPING MACHINES AND REAPER-BINDERS

Systematic attempts to develop horse-drawn reapers began in England and Scotland during the late eighteenth century. The patent for such a machine was granted in 1799, and this was followed by a spate of inventions, culminating in the much-praised reaper developed by the Scot, Patrick Bell, in the late 1820s.[32] To some extent, these English and Scottish experiments were paralleled in Ireland. In 1806, the *Belfast Commercial Chronicle* published a letter describing the demonstration of a machine at Moira, County Down, which the correspondent hailed as a success:

> To the Farming Society of the county of Down
> Gentlemen – I was so very fortunate a few days since, as to be present at the Rev. Mr. MacMullan's, Curate of Moira, where a new invented machine for cutting down standing grain was exhibited; as experiment, the constructor cut, in a field of oats, against seven of the best reapers in the field; the result was, on his ridge was double the quantity of sheafs as on theirs, and one over – his stubble was more level and two inches shorter than theirs; in fact, the unanimous opinion of at least sixty persons who were present (most of whom were farmers) was, that the machine effects the work of fifteen reapers, and its execution superior. Too much praise, I think, cannot be given to Mr Jellett, both for the invention and his condescension in exhibiting this machine where requested, as also his willingness to instruct others in its use, and to have similar ones made for any person wishing to have them, under his direction, by a mechanic in Moira, who he means to instruct for general accommodation … A FARMER.[33]

Unfortunately, like many contemporary reaper designs, Mr Jellet's invention seems to have disappeared without trace. Much greater excitement was aroused by the appearance of two American reaping machines, McCormick's and Hussey's, at the Great Exhibition in London in 1851 (fig. 88). Similar excitement greeted the appearance of the machines in Ireland, although several commentators claimed that the earlier Bell reaper performed better at trials. A

88 McCormick's reaper, 1851.

Hussey-style reaper, designed by a Belfast man, Richard Robinson, was shown at trials held in Belfast in 1852, but lost to a machine of Bell's design.[34] In 1853, Hussey and Bell machines, made by Crossley of Beverley, Yorkshire, were exhibited at the Irish Industrial Exhibition in Dublin. Once again, however, Bell's machine was judged superior.[35]

Despite the opinions of early observers, it was the American machines that were successfully developed by manufacturers. The English firm of Burgess and Key improved McCormick's reaper in 1854 by the addition of an Archimedean screw which delivered the cut grain to one side of the machine. In 1859, these machines were being advertised in an Irish agricultural journal, with sales figures (presumably for both Britain and Ireland) and a warning that orders for the machines should be made early:

Numbers of Burgess and Key reapers sold

1856 50   **1857** 250   **1858** 700

Owing to the excessive demand for last season, many orders were left unexecuted. Burgess and Key, therefore, earnestly request that the orders for the ensuing season be sent as soon as possible.

The advertisement also quotes a testimonial from several gentlemen who were present at a trial of one of the machines at Clandeboye, County Down, on 26 August 1858. These gentlemen affirmed the effectiveness of the Burgess and Key machines on both 'a light crop of oats growing on a clean surface, and on a heavy crop with clover and grass growing thickly in the bottom.'[36]

Combined reaping and mowing machines were being manufactured by the late 1850s. These were powered by the large, heavy running wheels on their sides. The wheels were connected by gearing to a rod known as a 'pitman', which worked backwards and forwards, and this movement was transmitted to the cutting knife of the reaper. By the 1860s, some agents as well as manufacturers of reaping machines were established in Ireland. At the Spring Show of the Royal Dublin Society, in April 1863, W. O'Neill of the Agricultural Implement Depot in Athy exhibited 'patent prize combined reaping and mowing machines'. Several Dublin exhibitors also had reaping machines on their stands. In August of the same year even more stands belonging to Irish firms selling reapers were exhibited at Kilkenny, including that of Thomas McKenzie of Munster Agricultural House, Cork. It is impossible to say from reports however, which, if any, of the exhibitors were manufacturers or agents. The largest Irish manufacturer of reaping machines was Pierce of Wexford, who probably began producing them in the early 1860s.[37] By the late nineteenth century, the firm was producing a variety of machines, most of which could be used for either mowing hay, or reaping grain.

In Britain, it has been estimated that by 1871 there were about 40,000 reaping machines, cutting as much as 25 per cent of the total grain crop. With a change of horses, a reaping machine could cut eight to ten acres in a day. Eight to twelve men had to tie sheaves and stook them, if the saving in cutting time was to be maintained, but overall labour requirements were about 50% less than when scythes were used. However, although the Irish Industrial Exhibition catalogue estimated that as much as fifteen acres of grain could be cut by a reaping machine in a day, the editor was cautious about the immediate value of the machines for Irish farming. The probable cost of reaping fifteen acres by hand was compared to that for reaping a similar area by machine. It was concluded that using the machines could lead to a saving of 5s. 10d. per acre. This saving was judged important, but only 'where the extent of land under grain crops is considerable'.[38] The same saving would not be so significant for smaller Irish farmers, and when the price of the Burgess and Key machines quoted in 1859 (£42 10s.) is considered, the caution of Irish

1 Spadesmen in County Waterford, *c.*1824 (UFTM).

2 Old Irish long-beamed plough in use near Hillsborough in 1780 (W. Hincks).

3 Robert Berry ploughing in the Ulster Folk Museum. Dancer,
the white horse, is an Irish Draught.

4 Bob Crosbie planting oats with a horse-drawn seed
drill at Munlough, County Down in 1978.

5 Making potato drills at Muckross Traditional Farms, photo courtesy of Toddy Doyle.

6 Hugh Chambers of Moneydarragh, County Down, making potato drills with a Mourne plough, *c.*1980.

7 Joe Kane of Drumkeeran, County Fermanagh, spraying
potatoes with a knapsack sprayer in 1985.

8 Robert Berry spraying potatoes in the Ulster Folk and Transport Museum.

9 Building a hay rick in Muckross Traditional Farms, photo courtesy of Toddy Doyle.

10 Liam Corry and John Canning pulling flax in the Ulster American Folk Park, photograph courtesy of Liam Corry.

11 John Canning making a stook of oats in the Ulster American Folk Park, photo courtesy of Liam Corry.

12 The Connemara Horse, from D. Low, *Domesticated animals of the British Isles*
(Edinburgh, 1845).

13 Kerry cow, from Low, *Domesticated animals of the British Isles.*
(Edinburgh, 1845).

14 Kerry cattle at Muckross Traditional Farms (photo courtesy of T. Doyle).

15 *Droimeann* Irish Moiled cattle with white stripes on their backs in the Ulster Folk and Transport Museum.

16 Wicklow sheep, from Low, *Domesticated animals of the British Isles* (Edinburgh, 1845).

17 Kerry sheep, from Low, *Domesticated animals of the British Isles* (Edinburgh, 1845).

89 Reaper binder, pulled by a Ferguson tractor, *c*.1938, photo courtesy of Museum of English Rural Life.

agriculturalists becomes understandable. Reaping machines presented Irish farmers, on a larger scale, with the same qualified advantages as the scythe. Harvesting could be achieved much more quickly, a major advantage in the changeable Irish climate, but more workers were required for this shorter period. By the end of the nineteenth century, reaping machines were common in the major tillage areas in Ireland, but on the smallest farms, sickles and hooks remained efficient harvesting tools.

In 1887, the American firm, McCormick's, put a reaper-binder on to the market. This machine not only cut grain, but tied it into sheaves with string. The machines were very soon for sale in Ireland, some being advertised as 'specially designed for Irish crops; will handle the tallest and heaviest grain'.[39] However, despite the effectiveness of reaper-binder design, and the obvious saving of labour, the spread of the new machines seems to have been slow before the First World War (fig. 89). It was only then that the government's compulsory tillage policies encouraged some large farmers to invest in these expensive machines. Once purchased, however, some machines had a long working life. Even after tractors replaced horses, reapers and reaper-binders were still used, either being simply pulled behind the tractor, or modified to be powered directly from the engine.

SHEAVES, STOOKS AND STACKS

Before the widespread use of binding machines, sheaves were tied by hand, with 'corn-bands'. These bands could be made in many different ways, but a simple method, used widely in Ireland, was described and illustrated in *Stephens' Book of the Farm*: 'The corn-band ... is made by taking a handful of corn, dividing it into two parts, laying the corn-ends of the straw across each other and twisting them round ... the twist acting as a knot'.[40] The sheaf was set in the middle of the band, the ends of which were then pulled around it, and twisted together. These twisted ends were then pushed under the band, and acted as another knot. If sheaves were to be left in the field for any length of time, it was important to have the grain heads in the bands pointing downwards, to allow rain water to drain off. If water collected in the heads, after a short time the seeds could germinate and sprout. Workers responsible for binding were also warned not to place the bands too near the stubble ends of the sheaves as this made them stand less well in the stooks in which they were dried.

The construction of stooks varied with weather conditions, and also the type of grain being harvested. Sheaves of grain, particularly oats, which had been bound while damp, were sometimes left standing singly in the field in the same way as retted flax. This was known in some areas as 'gaitin' (fig. 90a). The bottom of the sheaf was spread out 'like a tent' to make it stand better. The stooks most commonly described by nineteenth-century agriculturalists, however, were made up of two rows of sheaves, whose tops leaned on one another (fig. 90b):

> The binder, having bound up the sheaves, places them ... resting upon their bases, and upon one another. Five pairs of these sheaves, when the crop is oats or barley, and six pairs when it is wheat, may be conveniently placed together in rows ... The whole is covered by two sheaves, the butt ends of which lie towards each other, with the other ends divided a little, and pulled down so as to defend the upright sheaves.[41]

(b)

90 (a) A gaitin' of oats; (b) a stook of oats or barley, from *Stephens' book of the farm*, 3 vols (Edinburgh, 1908), vol. 2 p. 187.

(a)

The two sheaves laid along the tops of the others were known as 'hooding' or 'head' sheaves. These protected the sheaves they covered from both rain and birds. Farmers were sometimes advised to remove the 'head' sheaves on fine days, and to turn the other sheaves around, so that the sides which had pointed in towards the centre of the stook would face the air. It was also recommended that where grain had been grown on ridges, the stooks should be built across the furrows. This would assist the circulation of air along the central line of the stocks. Several agriculturalists advised that stooks should be aligned so that their ends faced north-south, as this would promote even drying. One omission from standard nineteenth-century Irish farming texts is the method of stooking described in the 1908 edition *of Stephen's Book of the Farm* as 'pirling':

> A plan of stooking sometimes pursued in certain exposed districts of the west and south-west is to set up two pairs of sheaves, the one pair at right angles to the other … When set up, the tops of the four sheaves are tied together about 9 inches under the apex, by a few straws pulled out of the top.[42]

There is plenty of photographic evidence that oats in particular were stocked in this way in Ireland (fig. 91), and the technique is still sometimes used on the

91 Pirlins of oats in County Tyrone (photo: Rose Shaw Collection,
Ulster Folk and Transport Museum).

rare occasions that stooks are made, yet the only evidence found so far that it
may have been used in the nineteenth century is a comment in Stephens that
'pirling' was probably more common 'fifty years ago' (plate 11). Farmers who
still occasionally make these stooks point out that a major advantage is that
they can be easily moved, both during and after drying. Since the sheaves are
bound together, the whole stook can be lifted at once. This means that it can
be easily shifted on to drier stubble. Lifting also allows some of the grass
bound up with the sheaves to fall out.

Grain was left in stooks until the sheaves were judged to be dry. It was
claimed by several agriculturalists that sheaves of grain cut with a scythe or a
reaping machine were looser than those cut with hook or sickle, and so dried
faster. The length of time taken for drying, however, depended most on the
weather. Some Irish farmers, especially in the midlands and southern counties,
built sheaves into 'ricks', which were intermediate in size between stooks and
stacks, to allow complete drying.[43] However, farmers hoped to move stooks to
the relative safety of grain stacks as soon as possible. Most stacks in Ireland

were built on a circular base. The sheaves were placed in layers, the stubble ends pointing outwards, and the grain heads meeting at the centre. Part of the skill in making a stack was to ensure that the sheaves also sloped slightly upwards towards the centre. This meant that rain water seeping through the stack tended to drain outwards rather than stagnate around the grain heads, possibly leading to germination. Stacks were given conical tops, and these were often roughly thatched to protect the grain from rain, and tied down with ropes (often *súgán* ropes of twisted hay), to make the structure stable in windy weather.

Larger farms had a 'stack-yard' or 'haggard' in which the stacks were built (fig. 23), but on smaller farms an area in the corner of a yard or field was set aside for the purpose. Some of these smaller farmers made bases for their stacks, simply by laying furze or thorn-branches on the ground, but 'corn-stands' were also common. These held the stack off the ground and also discouraged large-scale invasion by rats and mice. Stands varied widely in construction. By the mid-nineteenth century, iron stands were being produced by some local foundries, while on some large farms elaborate stone bases were made. On the other hand, some smaller farmers constructed crude stone bases from large flat pieces of local stone. Stands of dressed stone, shaped to resemble mushrooms, were also common. The length of time the grain was stored in stacks depended on when threshing could be carried out, and which threshing technique was used, but once the stacks had been built, the worst worries and hardest labours of the grain harvest were over.

CHAPTER TWELVE

Threshing and Winnowing Grain

S EED AND THE STRAW, the most valuable parts of grain crops, are usually separated before either is used. In Ireland, some of the techniques used to do this were regarded as exotic and primitive by the eighteenth century. One such technique, used to obtain small quantities of seed, involved burning the straw and husks, leaving the grain scorched, but otherwise undamaged. Burning might mean setting a whole sheaf alight, and then sifting the parched seed from the ashes. Alternatively, only the ears might be set alight and then knocked off the sheaf using a stick. A third method involved cutting off the ears from a sheaf and then burning them in a heap. Burning was forbidden by an act passed in the Dublin parliament in 1634, but memories of bread made from burned grain (Irish: *loiscreán*) have persisted until very recently.[1]

One very simple way to remove seed from a sheaf was by lashing (fig. 92). This was still practised on some small farms as recently as the 1980s, if only a little grain seed was required, for example to feed chickens. The sheaf was held at the lower end and the heads beaten against a hard object, such as a stone or a chair. This dislodged the grain, which fell to the ground or on to a cloth or sheet. Thatchers liked to work with straw which had been lashed, as the technique left it relatively undamaged. Wheat and rye, especially, were lashed as a preparation for thatching.[2]

Flails were used for threshing until well into the twentieth century (fig. 93). The simplest form of flail was a single stick which was used to beat grain-seed out of sheaves. This technique was known in Ireland as 'scotching' or light threshing (Irish: *scoth-bhualadh*).[3] For many centuries, however, the most common type of flail (Irish: *súiste*) used in Ireland has been made from two sticks, of between three and four feet in length, tied together. One stick was held, and the other used to beat the seed from the ears of grain. The different parts of flails are known by many different names throughout Ireland. Some of these are listed below:

Handstaff: *collop, cólapan, lamhchrann*
Striker*: buailteán, bóilcín,* souple
Tying: *iall,* thong, gad, hanging, tug, hooden, mid-kipple.[4]

92 Lashing oats on the Aran Islands, *c.*1930, photo courtesy National Museum of Ireland.

93 Flailing oats on Tory Island, County Donegal *c.*1930 (Mason Collection).

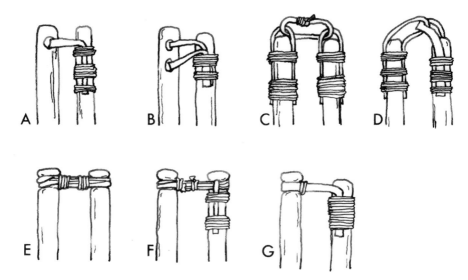

94 Common Irish flail tyings: (a) hole flail (Ulster); (b) double hole flail (Ulster); (c) cap flail (Leinster, South Ulster and East Munster); (d) sub-type of cap flail (Leinster and East Munster); (e) double loop flail (Munster, Connacht and West Ulster); (f) sub-type of double loop flail (Munster); (g) eye flail, drawing after Caoimhín Ó Danachair, 'The flail in Ireland', *Ethnologia Europaea,* vol. 4 (Arnhem, 1971).

The striker was usually made from a fairly hard wood, such as holly, blackthorn, whitethorn or ash, while the hand staff could be made from hazel, pine or oak. Ó Danachair has distinguished several types of Irish flails by differences in the way these sticks were joined (fig. 94).[5] The tyings, which could be made from several sorts of animal skin (eel-skin being especially valued), scutched flax fibre, tough flexible withies of willows or (less desirably), rope, show some regional differences.

Flailing could be carried out by a man working alone, or up to four men working together. If two or more men were threshing together, it was important that each worked at a constant, complementary rhythm. The hard knocks received by flailers when this was not done are still remembered with amusement. There was a basic difference in how flails were used, between those whose tyings allowed the striker to swing freely around the handstaff (fig. 94 e, f, g), and those where they could not do so (fig. 94 a, b, c, d). With the latter type of flail, the handstaff was turned as the flail was swung around.

A man working alone could thresh between two and four sheaves at once. The sheaves were laid at the worker's feet, with their heads laid facing one another. After a few strokes the sheaves were turned over, often using the end of the handstaff, and threshed again. There was some discussion amongst agriculturalists as to the best surface on which sheaves should be laid for flailing. In the early nineteenth century threshing often seems to have been carried out in the middle of a road, and threshing in the open air continued on many small farms until recently, although it has been claimed that in Leinster it was almost always done in a barn.[6]

Agriculturalists, and farmers who can still use flails, agree that a good threshing surface should 'bend to meet the flail'.[7] Arthur Young disapprovingly recorded threshing floors made of clay, or at Furness, County Kildare, a mixture of lime, sand, and coal ashes. The farmers who used these floors claimed that they did not damage the grain, but Young disagreed.

> The samples of Irish wheat are exceedingly damaged by clay floors; an English millar knows the moment he takes a sample in his hand if it came off a clay floor, and it is a deduction in the value. The floors should be of deal plank two inches thick, and laid on foists two or three feet from the ground, for a free current of air to preserve them from rotting.[8]

On some large Irish farms, barns were constructed to make flailing as efficient as possible, rafters for example being specially raised to allow the striker to swing freely. On most farms, however, whether threshing was carried out in the barn, in the farm kitchen, or in the open air, the main preparation was to place a threshing board beneath the sheaves. This might be a door unhinged from an outhouse, or a smooth board which had been specially made for the purpose. Farmers liked to work on boards about six feet square.[9] A large cloth was usually spread on top of this, to collect the grain beaten from the sheaves.

In 1812, Wakefield claimed that on small farms in many parts of Ulster, grain was threshed almost as soon as it had been harvested, because on such farms there was no room for a stack yard, or haggard.[10] Later in the nineteenth century, however, flailing was a task given to labourers, or male members of the farming family, during bad weather or after dark in winter. The advantages of the flail were listed by Martin Doyle as 'its simplicity, the power of giving employment to the labourers in the barn during wet days, and the convenience of having fresh straw for fodder every day'. Some attempts were made to calculate the amount a man could thresh in a day, using a flail. Doyle estimated that a good worker could thresh twenty stones of wheat, or thirty-

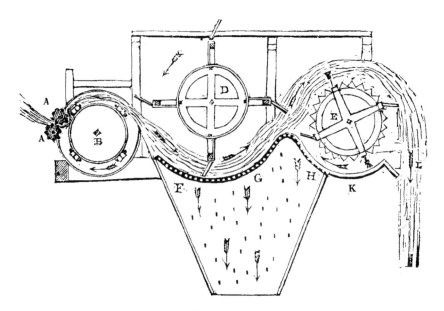

95 The internal workings of a threshing machine: (a) fluted rollers, through which
sheaves are fed into the machine; (b) cylinder with beaters attached to knock the seed
from the grain; (d, e) drums with rakes attached, to shake the straw, and pull it forward
through the machine; (f, g, h, k) the sparred bottom of the machine through which the
grain seed and chaff falls; (l) the opening through which the straw leaves the machine,
taken from John Sproule, *A treatise on agriculture* (Dublin, 1839), p. 77.

two stones of oats or barley per day. Threshing machines, however, according
to his calculations, could process more than twenty-five times this amount.[11]

EARLY THRESHING MACHINES

Most of the early developments in the design of machines for threshing grain
occurred in Scotland. In 1786, Andrew Meikle of East Lothian produced a
design which became basic to many later machines. Here the grain was
separated from the straw by a revolving cylinder, along which four bars of wood
faced with metal had been fixed at regular intervals. Water power was used to
turn the cylinder, or threshing drum as it became known, which was contained
inside a metal casing set around the drum, about one inch away from its
surface. Sheaves were fed into the drum, and the grain was beaten out by the
bars on the cylinder. In later designs, revolving rakes were added, to rake off
the straw. Meikle received an English patent for his machine in 1788, and from

96 A threshing machine and horse gears, manufactured by Sheridan of Dublin. The machine fixed to a wall inside the barn, and connected by a drive shaft to the horse gears outside. *Farmer's Gazette*, 1848, vol. 7, p. 398.

this time onwards threshers of his design began to be used on some large farms in Scotland and northern England (fig. 95).[12]

By 1800, threshing machines had been installed on some large Irish farms. A Mr Christy of Kirkassock, near Magheralin, County Down, claimed to be the first proprietor in Ireland to have installed a thresher. In 1796, he had inspected machines in Scotland and had drawings and a model made of one of these. A local workman used the plans to construct a similar machine on Mr Christy's farm. This early machine, and some others, were driven by horses (fig. 96). Mr Ward of Bangor, County Down, for example, had a 'horse engine', or gearing system around which the horses were driven in order to turn the thresher, set below the machine. This complex could be operated by two or three horses, and five workers. A boy drove the horses, while a man threw up the sheaves to be threshed to another worker, who unbound them. A fourth man fed the sheaves into the machine, and the fifth man removed the threshed straw. As with flails, threshing machines could provide inside work for labourers in bad weather.[13] Surveyors working for the Dublin Society reported that, by about 1800, threshing machines had been installed in several counties, including Cavan, Louth and Dublin. In the last county it was claimed that the machines were coming into use 'very fast'.[14] Machines continued to be installed on large farms during the first decade of the nineteenth century. By 1811, it was estimated that the cost of machines bought in Scotland and erected in Ulster by Scottish workmen had dropped to 50% of the 1800 price,[15] and during the same period, locally made machines were being erected in County

Cork. In 1813, for example, Mr James Stopford of Brookville farm gave the
following description of a machine installed on his farm:

> The thrashing machine … is, I believe a very good one, though it does
> not (at least as yet) perform as many combined operations as others of
> which I have read the description. It is worked by an overshot, or what
> is more properly called a balance, waterwheel … The person who
> prepared and put it up is Francis Barry of Brooklodge, near Riverstown
> who (although no more than 28 years of age, as he informs me) has
> already erected on his own account since the year 1805, in the contiguous
> and even some more remote counties, about forty threshing machines.[16]

During the early decades of the nineteenth century, threshing machines seem
to have been installed by farmers who were wealthy enough to take on the
expense of building a water-wheel, or a covered horse-engine. The Ordnance
Survey Memoirs recorded few threshing machines in the north of Ireland, even
by the mid-1830s. The situation reported from Lavagh Parish, County Cavan,
seems to have been typical: 'The flail and the centre of the road answer every
convenience of the threshing machine and barn'.[17] More widespread adoption
of mechanical threshing only came when smaller machines became available.

TYPES OF THRESHING MACHINE

Machines powered by water were usually large, and often formed part of a
complex which might also include a winnowing machine, and even corn
grinding stones. Although horses were used to power even the earliest
machines, agriculturalists advised against this. Horse gears seem to have put an
uneven strain on the animals, and the circular motion of turning them could
cause giddiness. Martin Doyle believed that a contrivance invented by the
Scot, Walter Samuel of Niddry, which meant that horses pulled with a
continuous and equal draught, would lessen the problem, but as late as 1863
Purdon's Practical Farmer described the use of horses for threshing as 'most
objectionable'.[18] Between two and six horses were used for driving the earliest
machines, but by the 1840s much lighter horse engines had been developed. In
1842, William Neill of the Hibernian Foundry in Belfast was advertising
himself as the inventor of a one horse-power threshing machine.[19] Smaller,
lighter horse-engines were a significant inducement to farmers to install
threshers. The new gearing systems made use of cast metal. The horses were

still harnessed to the outer end of a long arm (or arms), which turned around a central mechanism, but in the new machines the driving shaft that connected the horse-engine to the threshing machine came out from beneath the central gearing. In the most common arrangement, the driving shaft was often encased in a stonelined channel sunk in the ground. The machine operated by the new gearing systems also became increasingly standardized.

In 1837, the firm of Pierce began to make machines in Wexford, and soon became the largest manufacturer in Ireland.[20] By the 1860s, many smaller foundries were also making machines. In 1863, there were four manufacturers based in Belfast alone, but the northern foundry producing most machines was Kennedy's of Coleraine, which began making threshers in 1860. Between 1860 and 1940 about 2,000 machines were produced by the firm and installed, mostly in the north-east, but also in western counties such as Fermanagh and Donegal. Alan Gailey's detailed examination of the Kennedy papers shows how neighbours might influence one another when deciding whether to install a machine. In one example, a County Tyrone farmer, William Moorehead of Ballymullarty, Newtonstewart, wrote asking that Kennedy's should send a machine,

> as soon as you can as I would like to have her up before harvest. I expect she will be worthy of the praise you give your mills as I have some neighbours anxious to see her who wants one. You can let me know when to look for her at Victoria Station [Victoria Bridge] and when you will send the man to put her up.[21]

This particular letter also illustrates how the development of a rail network in the 1840s and 1850s allowed the transport of these heavy machines throughout the country.

Barn threshing machines were made in several sizes. The dimension usually quoted in advertiscments was the width of the threshing drum. Kennedy's at one time manufactured eleven different-sized drums, ranging from 18 inches to 54 inches in diameter. While many machines continued to be fitted with bars or beaters of the kind devised by Andrew Meikle, others had drums with rows of short iron pegs to beat out the grain. In 1849, both 'beater' and 'peg' drums were being manufactured in Cork.[22] In England it was claimed that 'beater' drums damaged the grain, although it was also argued that the damage was more the result of the speed at which machines were worked. In Ireland, on the other hand, there was some discussion as to whether or not 'peg' drums

97 Hand powered thresher in use.

broke up straw so much that it was useless for thatching. Threshing machines
with both types of drum continued to be made, but for larger machines the
beater drum seems to have been more commonly adopted.[23]

All threshing drums turned at speeds which could make them dangerous to
work with, and also mean that weak parts of badly constructed machines were
liable to break. The problem of damage to machines could be lessened by
building up the speed of turning gradually. After top speed had been reached
the corn should be fed in regularly, a rate of one sheaf at a time being sufficient
for most machines. Grain should not be put in to such an extent that it slowed
down the turning of the threshing drum. This was to be watched for
particularly if oats with long, damp straw were being threshed.[24]

Two developments meant that even very small farms were able to make use
of mechanized threshing technology; the manufacture of portable threshers,
and of very small machines, turned by hand.

Hand-operated threshing machines were known in Britain by the early
nineteenth century. By 1813, they had made their appearance in Ireland (fig. 97).

> The Dublin Society has lately got a threshing machine constructed to be
> worked by four men; two of them are employed in turning the machine;
> and the other two are engaged in feeding and carrying away the corn and
> straw. It is capable of threshing 20 barrels of oats per day or 16 of wheat.
> It can easily be removed from one place to another, and can be worked
> in a room 12 feet square.[25]

98 Steam powered thresher, exhibited in Dublin in 1853, and at the Dundalk Show in 1859, *Agricultural Review*, 1859, vol. 1, p. xcviii.

In 1849 *The Irish Farmer's Gazette* enthusiastically reported the exhibition of another hand-threshing machine in the Royal Dublin Society's agricultural museum, as a new solution to the problem of the expense of buying a thresher. The rate of working of this machine, twelve barrels of oats, and eleven barrels of wheat per day was less than that claimed for the earlier machine, but the makers, Barrets of Reading, Yorkshire, supplied many machines to Irish farmers during the second half of the nineteenth century. Machines which could be worked by pedals as well as being turned by hand also became available, and early in the twentieth century some of these were supplied to small farmers in those western counties known as the 'congested districts'. These machines cost only £7 each, and if five men were working them, they could thresh between twenty-two and twenty-four stones per hour.[26]

The other solution widely adopted to overcome the problem of the high purchase cost of barn threshers, was to hire a portable threshing machine, powered using a steam engine, horses, a stationary engine, or a tractor (fig. 98). The development of lighter cast-iron gearing systems meant that it became possible to move the entire threshing system from farm to farm. A horse-operated portable thresher was patented in England in the 1840s, and shown at the Great Exhibition of 1851 in London. In 1853, similar machines were shown at the Irish Industrial Exhibition in Dublin. John Sproule expressed some concern that this type of machine would be prone to breakages during

movement. He accepted however, that well-made machines, such as that
exhibited by Ransomes and Sims of Ipswich, would be strong enough to take
this strain. He also pointed out the potential value of portable machines for
small farmers, as a machine hired for a few days could thresh all the grain
produced on the farm.[27]

In the 1860s portable threshing machines were claimed to be 'partially' used
in Ireland, and horse-operated portable threshers continued in use well into
the twentieth century.[28] Portable machines became much more common when
steam power was applied to threshing. A Lincolnshire firm, Tuxfords of
Buxton, were claimed to be the first to develop this combination, in 1842.
Another English maker, Garrett of Suffolk, exhibited a portable steam
threshing complex in Dublin in 1853.[29] A machine made by this company was
successfully tested on the farm of a Mr Christy of Adare, County Tipperary, in
1859,[30] and during the next eighty years the machines became increasingly
popular. One effect of their increasing use was probably to lower sales of fixed
equipment, especially after the First World War.[31]

Barley presented particular problems during threshing. The heads of this
crop have long spines or 'awns', which could not be removed efficiently by
either flails or most unmodified threshing machines. Removing the awns could
require a separate operation, which became known as hummelling, awning, or
avelling.[32] Implements known as 'hummellers' were devised for this. One
common type had a shaft with a T-shaped handle, and at the other end a
square metal plate made up of parallel strips of iron. The barley was spread on
the ground and hit with the hummeller, until the awns were chopped off.

Threshing machines could be modified so that they hummelled barley as
part of the threshing process. Fluted covers were manufactured which could be
fitted over the threshing drum. In 1839 the *Irish Farmer's Gazette* suggested that
farmers could do this themselves by replacing the cover of the threshing drum
with one which they perforated with small holes. The barley awns would be
taken off as they were rubbed against the holes. In the same year, however, a
peg-drum threshing machine made in Cork was claimed to remove barley
awns without any modification being necessary. This was said to be due to the
velocity at which the machine was driven. As we have seen, however, very high
speeds of working were declared unsafe by many agriculturalists.[33]

The major advantage claimed for threshing machines was that they were
fast. One of the more unusual benefits claimed as a result of this speed was

that it 'saves pilfering, [the grain] … not being kept too long under the hands of workmen'.[34] However, although the same writer claimed that threshing machines were much more effective than flails in beating grain out of sheaves, their overall efficiency was not unquestioned. In 1813 the *Munster Farmer's Magazine* advised that unless sheaves had been cut neatly with a sickle, so that they could be fed into a machine very evenly, each sheaf should be threshed twice. Also, sheaves with very long straw could slow the machine down, and agriculturalists advised that long straw should be cut before threshing. This had to be done carefully, however, since if the straw was cut too short the sheaves might pass through the machine too quickly to be properly threshed.[35]

Worries were also expressed about the general efficiency of machines' construction. In 1854, John Sproule wrote,

> Year after year we … [have found], that even the best manufacturers continued to turn out machines, the three-fourths of the power to work which was necessary for merely putting them in motion. Thus the four horse power threshing machines actually required three horses to work them when empty, leaving to the fourth the whole labour of threshing the corn.[36]

The calculations of costs involved in threshing with either flails or threshing machines were complex. The earliest machines were large and expensive, whereas Irish farm labour was so cheap that even in the mid-nineteenth century it was not always clear whether it was worthwhile for an affluent farmer to purchase a threshing machine. In 1849, for example, a correspondent in the *Irish Farmer's Gazette* asked if it was more economical to purchase a threshing machine than to have oats threshed with flails, which he could have carried out for the cost of 3½ *d.* per barrel of seed produced. The *Gazette* replied that it would certainly be much cheaper to thresh with a machine. This was qualified, however, by some remarks on the proper use of labour. Threshing machines and flails were seen as useful means by which a farmer could keep labourers working during bad weather, or after dark. This was usually presented in terms of the farmer getting maximum returns from his labourers, especially if they were hired servants living on the farm. The *Gazette*, however, echoed the call of English farm workers who opposed the introduction of threshing machines in that country: 'In order to give employment, we would prefer thrashing [with flails] at 3½ *d.* a barrel than using a machine … at the same time we are

99 Portable threshing machine, powered by a Ferguson tractor *c.*1938, photo courtesy of the English Museum of Rural Life.

under the impression that 3½ *d.* a barrel is very poor hire'.[37] The declining number of farm labourers in the late nineteenth century, and the methods just described that allowed even small farmers could avail themselves of the services of a threshing machine, meant that the latter became increasingly dominant. However, once again it is clear that in comparing 'common' and 'improved' techniques, we cannot simply oppose them to one another as efficient and inefficient.

One unintended consequence of hiring portable threshing machines, was that ties of mutual help between neighbours were strengthened (fig. 99). Mechanization has often been cited as a cause of the decline of neighbourly help, but the number of people required to operate a threshing complex (as many as fourteen), meant that many farmers had to recruit neighbours to help them, and during the threshing period, in autumn, teams of neighbours would

100 Winnowing grain on the Aran Islands in the 1930s, photo courtesy of the National Museum of Ireland.

move from farm to farm helping with the threshing. These work groups were known by many names including a *meitheal*, gathering, or fiddler, and working with a threshing group is remembered as a particularly happy experience. Mrs Mary Wilson, of Townavanney, County Fermanagh, summed up her own memories of threshing days in the early twentieth century.

> It was a time of really thanksgiving … Everyone helped each other and they were all very good and kind … and even they would stay, you know, on after dark at night, to help finish up … In fact, it was the top item of the year … that you had the corn safely gathered.[38]

WINNOWING

After threshing, the husks or shells around grain seed were removed. For most of the eighteenth century, and indeed into the twentieth, this was done out of doors on a breezy day by simply pouring grain from a container (fig. 100). The heavy grain fell straight down on to a sheet laid out for the purpose, while the light husks were blown away. This open air winnowing could take place either on a road or in a field. Arthur Young noted instances of both, as did William Tighe in County Kilkenny in 1802.[39] In early nineteenth-century Ulster, winnowing was claimed to be women's work.[40] High windy places were particularly valued as winnowing places. These were sometimes known as 'Shilling' (shelling) hills.[41]

The grain to be winnowed was shaken from riddles, sieves or circular trays. Riddles were also used to clean grain before winnowing. During the nineteenth and early twentieth centuries, local experts made riddles from wooden strips (usually of ash), woven and attached to a circular frame. The size of the meshes or holes in the riddles was varied according to what type of grain was to be cleaned. One maker, whose techniques were recorded in County Dublin during the 1940s, made riddles with meshes of the following dimensions:

> Barn riddle – meshes 25 x 32 mms.
> Oats riddle – meshes 6 x 9 mms.
> Barley riddle – meshes 10 x 6 mms.

This craftsman also made winnowing trays, known as 'blind sieves', since the wooden strips were closely woven together so that there was no space between them.[42] The Irish name for this type of tray, *dallán*, was also given to winnowing trays made from sheep or goat skin stretched around a circular wooden frame. Skin trays like these were found throughout Ireland. On some western islands where wood was scarce, the rims were made from rings of straw rope (Irish: *súgán*) and some trays made entirely of coiled straw rope, were used in Donegal.[43] The names given to the trays varied. In the north a smaller version of the *dallán* was known as a 'wight' or 'wecht', while in the south the term *bodhran* or 'borrane' was used. *Bodhran* is also the name given to a one-sided Irish drum which became popular with Irish folk musicians during the 1960s. It seems probable that before the modern musical development of the bodhran, it had the dual functions of tray and drum.[44]

101 Barn fan, illustrated in *Stephens Book of the Farm,* 3rd ed. (Edinburgh, 1871).

BARN FANS

James Meikle, the father of Andrew Meikle whose work on developing threshing machines was mentioned earlier, developed a machine for winnowing grain, based on designs he had seen in Holland. Meikle's winnowing machine or 'barn fan' set up a current of air by turning canvas sails. As grain was fed into the machine, it fell through this artificial breeze which separated the seed from the husks. The seed was also cleaned by passing through a series of sieves. In 1770, the Englishman James Sharp constructed a more compact version of the machine. Here, the draught was created by wooden fans, turned by a handle on the side of the machine's wooden casing (fig. 101).[45] The simple construction of these machines made them cheaper than threshing machines, and it has been suggested that barn fans were adopted in Ireland a generation or more before threshers. By the beginning of the nineteenth century, barn fans were claimed to be in common use around Dundalk, for example.[46]

The principle behind the construction of barn fans was was easily grasped. This was illustrated by the Revd Gerald Fitzgerald of the parish of Ardstraw, County Tyrone:

> Soon after I came to this parish, in 1806, I got a winnowing machine, and having on a calm day, during its operation in the barn, asked a man who used before to winnow in the field, how he liked it? He exultingly replied, 'O sir, we need not now be looking abroad for the wind, we have her under lock and key!'[47]

Barn fans were also easy to use. Because of the gearing system, the fans could be turned for a long time without difficulty. By the mid-nineteenth century fans not only separated the grain from sticks and stones, but also delivered grain of different weights, and therefore different quality, from different spouts. Wire-meshed riddles of different sizes were fitted to the machine depending on which type of grain was being winnowed. Wheat riddles had five meshes to the inch, barley riddles four meshes to the inch, and oats riddles three meshes to the inch. A well-made machine would winnow or 'dress' oats in one go, but farmers were advised that wheat and barley grain should always be put through the machine twice.[48]

In large early nineteenth-century complexes, winnowing machines were commonly placed on the lower floor of a barn, with a threshing machine placed on the floor directly above. This meant that newly threshed grain was easily passed on for winnowing. By the 1830s, however, American manufacturers were producing machines which threshed and winnowed grain as parts of a single operation. Steam powered threshing machines such as Garrett's also bagged the grain.[49] Hand-threshers and barn-threshing machines usually did not perform these extra functions, however, and on farms on which these were used, winnowing remained a separate activity.

GRAIN STORAGE

Some agriculturalists advised farmers not to thresh their entire grain crop at once. In 1849, the *Irish Farmer's Gazette* compared the losses of grain kept unthreshed in stacks, and that stored in a heap after being threshed. The *Gazette* emphasised that the amount of grain lost from a stack depended very much on the dryness of the site on which it had been built, but estimated that the loss in general would be between 10 and 20%. However, it was claimed that this was a small amount compared to that lost from threshed grain. In a stack, the *Gazette* argued, 'the grain will be kept sweet, if properly saved, and the only loss will arise from evaporation; but if thrashed, and kept over, the corn is very rarely kept sweet … so that it is a practice that should not be adopted'.[50] We have seen, however, that Irish farmers, whether depending on flails or large threshing machines, tended to thresh their grain crops as quickly as possible. Where grain was grown for cash, the crop was often sent to market almost immediately.[51] Grain-seed kept on the farm, however, was stored in a variety of ways. In 1802, the lack of proper storage on some County Kilkenny farms was severely criticized. 'In unfavourable weather corn sometimes lies exposed in a hovel called a barn, to vermin and damp'.[52] Field-workers in County Cork have recorded an assertion that grain which was stored unwinnowed was particularly vulnerable to attacks by rats and mice. The husks of the grain made the mass more compact, allowing them to tunnel into the heap. Dry, winnowed grain formed a much more mobile mass, and could not be tunnelled into so easily.[53]

The experts of the *Farmer's Gazette* who advised against early threshing may have had in mind heaps of unwinnowed grain kept in the way observed and condemned by Tighe. This quality of storage must be contrasted with other evidence, however. Arthur Young, for example, gave high praise to a granary he examined on one large farm in County Antrim: 'Mr Lesly's granary is one of the best contrived I have seen in Ireland; it is raised over the threshing floor of his barn, and the floor of it is a hair-cloth for the air to pass through the heap, which is a good contrivance'.[54] Some small farmers also stored grain carefully. One method, recorded in Counties Cork and Kerry, involved the construction of a cylindrical granary using very thick coils of straw rope (Irish: *súgán*). The base of the granary could vary from between 6 feet and 26 feet in diameter. Circular stands made from slabs of stone were sometimes used to hold the base off the ground, but in any case a protective layer of furze, twigs or brambles was laid down before being covered by a layer of straw. The sides were built up from coils of straw rope, grain being added as the work progressed. Recorded examples had sides between 5 and 6 feet high. The construction was completed with a domed roof of thatch which brought the overall height to between 7 and 12 feet. A.T. Lucas has argued that this type of granary (known by several names, including *fóirin, síogog, sop, dárnaol* and *lochtán*) has been made in Ireland for at least several hundred years, and probably once had a much wider distribution throughout the country.[55] Since the construction was filled as it was being built, this may have provided a further incentive for farmers to thresh their crops quickly.

Threshing, winnowing and storing grain crops were closely related operations. The method that a farmer used depended on how quickly he wanted grain, his ability to gain access to technological developments, and how he decided to use available labour. It is only by taking all these considerations into account that we can understand not only how single operations were performed, but also how these were related to other aspects of the processing of harvested grain.

CHAPTER THIRTEEN

Horses and Ponies

DURING THE MIDDLE AGES, horses supplanted oxen as the principal draught animal in Ireland. Horse-powered technology was a major element in improved agriculture in the nineteenth century and the horse remained the main source of draught power on Irish farms well into the twentieth century. It was only at the end of the 1950s that tractors became more common.

BREEDS OF HORSE

The quality of Irish horses had been recognized for centuries[1], but by the early nineteenth both horses and their use by farmers were often criticized by observers. Wakefield, writing in 1812 commented, 'The condition of the working horse in Ireland is altogether miserable … Bad feeding and hard working keep them in a state of wretchedness hardly to be conceived'.[2] Although the origins of indigenous breeds of Irish horse are obscure, early Irish texts distinguish between two main categories of horse used by farmers, for riding, and for working.[3] Early Irish writings do not allow us to attribute precise physical characteristics to particular types of horse. However, skeletal remains from archaeological evidence suggest a small type of domesticated horse which stood thirteen to fourteen hands high and had strong Arab features. It has been argued that these small horses or ponies were the descendants of a form of wild horse, *Equus Caballus Celticus* and the ancestors of surviving breeds of native Irish horses (fig. 102). During the Middle Ages a small light Irish horse called the 'Hobby' was known and exported throughout Europe.

The Hobby was valued as a light cavalry horse,[4] but its stamina and mettle gained it high repute as both a saddle and pack animal. Some writers have argued that the Hobby was the foundation stock of more recent breeds and types of Irish light horse and pony. A shared characteristic between Hobbies and more recent native equines used by Irish farmers was their ability to perform a variety of tasks, doubling up as a riding horse and work horse.[5] In early literature small Irish work horses or ponies are often referred to as *gearrán*. In the later Middle Ages the term passed into English in the form 'garron'[6] and it occurs frequently in more recent written sources (fig. 103).

102 *Equus caballus celticus,* identified by J.C. Ewart, taken from R. Wallace, *Farm livestock of Great Britain* (Edinburgh, 1923), plate 133.

Early nineteenth-century descriptions of native garrons, although not always complimentary, often attribute similar characteristics of size, hardiness, and agility. In 1804 Sir Charles Coote's account of Irish garrons in County Armagh describes them as having slender bones, remarkable for speed and hardiness, and as being reared in the mountainous districts.[7] In 1814 the Revd George Sampson remarked:

> The strains of horses distinguishable in this county (Derry) are first the native garron of the mountainous country; these are thinly made, in general have crooked hams seldom exceed 14 hands, are often much lower; the prevalent colours are bay and sorrel. This breed have, for the most part, a gentle head and aspect, with nice shanks,[8]

In 1812, the Revd John Dubourdieu described similar small horses in County Antrim.

103 Garrons at Banbridge, 1783 (W. Hincks).

There is a very hardy, strong, though small, race of horses … much in use on the northern and north-eastern coast, and in the mountains. They are active and sure-footed, but few of them exceed fourteen hands high, and many are much lower. In shape, their defects are, want of height and length before, and, behind, their hams approach too close; but their backs and limbs are excellent, and their paces far above what would naturally be expected from their apparent strength, being equal to support a journey of equal length with a horse double their bulk, when not unmercifully loaded.[9]

As well as the shared characteristics of size, hardiness and agility, the above nineteenth-century descriptions suggest that small native horses were found mainly in marginal, mountainous areas. In 1901, the Department of Agriculture and Technical Instruction assessed the Irish horse breeding industry, and identified two principal areas which possessed small horses 'deserving of special mention'. These were the moorlands of western Galway and Mayo, home of the Connemara pony, and the mountainous districts of north Antrim which produced the Cushendall pony.[10]

Connemara Ponies

Hely Dutton, in 1824 claimed that Connemara 'has long been famed for its breed of small hardy horses' (plate 12).[11] Although native horses of particular types possessed similar characteristics, their breeding was carried out by farmers at an informal level and based on local and individual needs. As a result horses from a particular area, as well as sharing features, often had a wide diversity of characteristics within a type. The Connemara was such a case. J.E. Ewart's study, commissioned by the Irish Department of Agriculture in 1900 to assess the potential of the Connemara for improvement, showed that due to the absence of a selective breeding programme, and years of uncontrolled cross breeding or 'blending' of native horses with introduced animals, there was not a 'single distinct strain' of Connemara pony. Instead he estimated that there were five categories or types of Connemara which he referred to as the Andalusian, the Eastern, the Cashel, the Clydesdale and the Clifden. Ewart concluded that of the five types the Andalusian and the Clifden were of particular interest (fig. 104). The Andalusian he claimed, represented the original or old breed, the Clifden 'the best kind of the Connemara … well worth preserving, not only because it was well adapted for the country, but also because it would be invaluable for crossing with other breeds'. The Andalusian was described as standing '12 to 13 ands high; some are black others are grey or chestnut but the most characteristic specimens are of a yellow dun colour'. Although only slightly higher than the Andalusian, standing 13.2 hands at the withers, Ewart described the typical Clifden as being heavier and stouter, and very different in build. He suggested that these differences in confirmation might either have resulted either from cross breeding, or from the Andalusian and the Clifden having different aboriginal ancestors. While Ewart saw the need for the standardization of type through a breeding programme based on selected stallions and mares, he also urged that any attempts at improvement must not destroy the essential qualities of hardiness, stamina and docility inherent in the native Connemara pony. These qualities, he believed, were shaped by the natural and working environment within which the Connemara had to survive.[12] However, it was not until 1923 that Ewart's recommendations were acted upon with the setting up of the Connemara Pony Breeders' Society, and the opening of the Connemara Stud Book. As a result, the ponies have been widely exported throughout the world.

104 Andalusian type Connemara pony, *c.*1910 (Welch Collection, Ulster Museum).

Cushendall Ponies

Like the Connemara, the recently extinct Cushendall pony was a product of its natural and working environment (fig. 105). In 1904, the Department of Agriculture noted the Cushendall and Connemara ponies as being striking examples of how 'horses become modified through adapting themselves to their surroundings' and 'admirably calculated to live and thrive and do useful work under the circumstances and amidst the surroundings in which they have been evolved.'[13] As with the Connemara, local farmers in north County Antrim bred the Cushendall pony at an informal level and to suit local needs, crossing native ponies with imported animals to varying degrees of success. In the nineteenth century, this included ponies from the Highlands of Scotland and in the twentieth century, bloodstock, Shires and Irish Draught horses. No official stud book was ever opened or breeder's society formed for the Cushendall pony. However, despite the lack of a selective breeding programme, a typical Cushendall was said to have been small in stature, standing around fourteen hands high, with an Arab head and possessing the essential qualities of

105 Cushendall ponies *c.*1906 (F.J. Bigger Collection, Ulster Museum).

hardiness, surefootedness and strength. Antrim farmers claimed that the
demise of the Cushendall was due to the decrease in tillage and spread of the
tractor after the Second World War.[14] While these are major factors
contributing to the end of its use by local farmers, additional factors, related
to the implementation of a selective breeding programme and its role as a
work horse also contributed towards the disappearance of the Cushendall.

Kerry Bog Ponies
The value of a selective breeding programme and officially recognized stud
book in helping to conserve breeds is aptly illustrated by the recent rescue of
the Kerry Bog pony from extinction (fig. 106). Referred to by locals as the 'old
pony', and reputed to have been taken in large numbers by the British army to
serve as a pack animal during the Peninsular wars, the Kerry Bog pony is
described as ten to eleven hands high, with a dished Arab-like face, chestnut
coloured with a flaxen mane and tail, or liver chestnut with a black mane and
bay or grey tail, and having a docile temperament. By the early 1990s it was
said to be on the edge of extinction, but the dedicated efforts of John Mulvihill

106 Postcard of Kerry Bog Pony, photo courtesy John Mulvihill.

from Glenbeigh, County Kerry, changed the fortunes of the pony. After chance discovery of a stallion and locating a suitable mare, Mulvihill determined to save the Kerry Bog pony by establishing a stud farm. Part of his strategy to achieve this, and conserve the pony as a distinct breed, included blood typing to identify genetic markers for future breeding, official recognition by registering the breed with the Irish Horse Board, the opening of a stud book to record lines of breeding and the formation of the Kerry Bog Pony Society. John Mulvihill has been successful in implementing all of these elements of his strategy and at present the Kerry Bog pony looks safe from extinction.[15] However, another important factor of Mulvihill's strategy to save the pony involved a significant change from its working role on the farms and bogs of County Kerry, which will be discussed below.

Irish Draught Horses
Ireland's most famous breed of large working horse is the Irish Draught (fig. 107). The first 'authentic reference' to the Irish Draught horse is reputed to come from the end of the eighteenth century, when increased tillage also

107 Toddy Doyle with an Irish Draught horse from Muckross Traditional Farms, photo courtesy of T. Doyle.

increased demand for 'horse labour' among Irish farmers.[16] Heavy horses from England and Scotland were imported, but it was found they did not suit the needs of the majority of Irish farmers, who required a hardy multipurpose animal. The type of horse best suited to Irish farming practices of the period evolved through the selection and breeding of heavier types of native horse by the farmers themselves. The horse produced was said to be a 'medium sized farm horse of excellent quality' capable of a range of work and described as rarely exceeding 15.3 or 16 hands high, with strong lean legs, plenty of bone and substance, short back, strong loins and quarters and having a strong neck and smallish head.[17] As well as being a versatile farm horse, a valuable characteristic of the breed was the suitability of mares for crossing with thoroughbred stallions to produce the famous Irish Hunter.

During the nineteenth century, breeding of Irish Draught horses remained at a local level and in the hands of individual farmers. A depression in the agricultural sector and an accompanying decrease in tillage during the latter half of the century led to a decline in the number of 'typical' Irish Draught horses. This, coupled with the informal system of breeding and the resulting progeny, became a matter of concern. In 1901, the Department of Agriculture and Technical Instruction, assessing the potential for the development of the hunter breeding industry in Ireland noted,

> Quite a large percentage of the mares by which Irish hunters are produced are the property of small farmers … Unfortunately for the country, the breeding of these 'old Irish' mares has not hitherto received the attention which it merited. Numbers of them have a dash of thoroughbred blood in them but the majority are got by sires of such mixed breeding that from the standpoint of pedigree they are but mere mongrels.[18]

The improvement of the Irish Draught was considered a question of national concern, and its restoration vitally important for the future of the horse-breeding industry of the country.[19] Raising Irish Draught horses to a standard which suited both the needs of the small farmer and hunter breeder, required a move from an informal system of breeding operated at the level of individual farmers, to a formally organized system of selective breeding controlled at national or government level. In 1905, the Department of Agriculture initiated a special horse improvement scheme, under which subsidies of £50 per annum were offered 'to owners of the old Irish Draught type and of half-bred stallions of the hunter type.'[20] Due to the small number of stallions originally selected, breeders were often required to bring their mares long distances to be served, and as a consequence the scheme was discontinued in 1915.[21] However, although only moderately successful, the scheme did produce a number of animals of desirable quality for further breeding and also showed that there was sufficient stock available to firmly establish the breed.

The impetus to achieve this came, not at national level, but at international level and involved a divergence in the roles for the Irish Draught. With the onslaught of the First World War, tillage in Ireland increased and the demand for strong active draught horses grew. The demand, however, was not restricted solely to the sphere of agriculture. The British army also had an urgent need for strong active draught horses for field artillery purposes. It was recognized that the versatile Irish Draught horse was well suited to both the needs of Irish farmers and the army, and once more attention was focused at government level on its development. In 1917, the Department of Agriculture opened a Register for Irish Draught Horses and in that year department inspectors had 372 mares and 44 stallions for registration. By 1938, the number of registered stallions had risen to 183. However, the spread of the tractor after the Second World War heralded the demise of the working farm horse and this was reflected in the number of registered Irish Draught stallions, which fell from 187 in 1944 to 63 in 1978.[22]

Despite attempts to introduce heavy horses such as Clydesdales and Shires, their use on Irish farms was limited. In terms of improving native animals through interbreeding, the resulting crosses, as mentioned earlier, were found to be unsatisfactory and their adoption as a working farm horse was restricted to a few areas. Early nineteenth-century observers argued that the generally small size of Irish farms, and the increased costs of upkeep in terms of feeding, made the use of heavy horses an uneconomic proposition for the average tenant farmer. At the start of the twentieth century, the counties noted for heavy horses were Dublin, Louth, Antrim, Down and Derry.[23] In recent times, heavy horses, particularly Clydesdales, have enjoyed a revival of interest amongst farmers in the north of Ireland. However, the revival has involved a significant change of role for heavy horses.

CHANGING USES

In 1874, Thomas Baldwin estimated that there were upwards of 500,000 horses in Ireland, two-thirds of which were used for agricultural purposes.[24] Irish breeds of horses that could perform a wide range of tasks and survive in harsh conditions survived as working farm horses from early times well into the twentieth century. Commenting on the jobs carried out by the Connemara pony, J.E. Ewart describes the pony working variously as a pack horse, riding horse and cart horse and using an array of local farm implements.

> Without a pony the peasant farmer in the west of Galway is all but helpless … The work of the ponies varies with the season of the year. At one time they may be seen climbing steep hillsides heavily laden with seaweed, seed corn, or potatoes; at another time they convey the produce to market. Sometimes it is a load of turf, oats, or barley; at other times creels crowded with a lively family of young pigs. Returning from market each pony generally carries two men, one in front and the other on the pillion behind … In Clifden and other centres, as on the larger holdings and some of the small farms close to the main roads, cars, turf, and other carts take the place of the pack-saddle and pillion.[25]

However, the continued survival of the Connemara has meant a change in its roles, from a working animal to a riding animal, exported and often bred abroad. Ironically, the successful export of the modern Connemara pony has caused a decline in the numbers of brood mares at home and is a source of worry for those authorities who are afraid that the Connemara will cease to be 'distinctively Irish'. This modern concern supports Ewart's earlier view that the harsh conditions of its native environment shaped the essential qualities of the Connemara pony.

The Cushendall pony, like the Connemara, was a hardy, multipurpose animal capable of performing a range of tasks. Local farmers from the Glens of Antrim who worked with the Cushendall said it was 'an all round pony, you could plough with him, cart with him, or if you wanted a bit of pleasure, you could put him in a trap and drive'.[26] The Cushendall was a good size for pulling small, locally-made slide cars and wheel cars, and was also said to be more manouverable in the small sloping fields of the Glens than the larger Clydesdale or Shire, when using 'improved' implements and techniques such as swing ploughs for lea ploughing and drill culture for potato cultivation. While the spread of the tractor and a decrease in tillage were major factors which contributed towards its disappearance, the failure to develop the versatile characteristics of the Cushendall pony through a selective breeding programme, and the establishment of an alternative role for it, such as that of a high-class riding pony, also contributed to its demise.

Known as the 'maid of all work', the Kerry Bog pony was also a hardy, multipurpose animal used for a range of work on the small farms of the south west. In recent times it has been particularly valued for its work in the bog, where it was used for bringing home turf for domestic fuel. One method of transporting the turf was in creels placed on simply constructed wheel-less slides, with shafts made from holly or birch trees. Later, small wheeled carts came into use, particularly in less boggy areas, although the pony was also used with improved forms of transport.[27] However, an important factor in the continued survival of the Kerry Bog pony is not its re-emergence as a work horse. John Mulvihill's and the Kerry Bog Pony Society's ultimate aim is to set up a 'heritage bog workshop', housing a range of crafts and craft workers, with the Kerry Bog pony as 'central character and symbol.' The pony is no longer valued as a working animal but has become a heritage icon.[28]

The Irish Draught horse was also a hardy multipurpose animal, able to survive bad weather with relatively little feed. It was used on small farms 'for every class of work ... for ploughing or harrowing one day, for hauling heavy loads of farm produce the next, and on the third, perhaps, for driving to market at an eight or nine miles an hour trot'.[29] However, the Irish Draught has not survived because of its usefulness as a general purpose farm horse, but because of its role as the foundation stock for the Irish Hunter. Conversely, the demise of the Irish Draught as a working horse has had implications for its continued survival as the foundation stock of the Hunter. The loss of its direct economic role as a working farm horse, coupled with the growing demand for half-bred horses for competitive and leisure riding has made it more profitable for Irish farmers to cross Irish Draught mares with thoroughbred stallions rather than Irish Draught stallions.[30] At the end of the twentieth century, authorities worried by this breeding trend, and the possible extinction of the foundation stock of Irish Draught mares, set up various bodies and schemes to encourage farmers to mate Irish Draught mares with Irish Draught stallions.

As mentioned earlier, the use of introduced heavy horses such as Clydesdales and Shires was limited mainly to the northern and eastern parts of Ireland. They were kept on large farms in districts where heavier tillage was practised, and near cities, for use in heavy haulage. This distribution pattern was reinforced in the early twentieth century by the Department of Agriculture, which limited the areas in which Clydesdales and Shires could be registered to the 'province of Ulster, the counties Louth and Dublin, and the district within a radius of ten miles of Cork city'.[31] As with other working horses, the use of heavy horses for farm work declined with the increased use of the tractor.

In recent years, there has been renewed interest in working horses at agricultural shows and events. One type of event which has enjoyed a significant revival among the farming community is the horse ploughing match. As we have seen, these became widespread during the early nineteenth century, when they were regarded by landlords and farming societies as a means for the encouragement of new farming techniques by promoting the use of improved plough types and tillage practices through competition. It can be argued that the revival of horse ploughing matches has a symbolic significance in which the horse plays a central role. The horses are a public attraction in themselves at matches, and it is especially their use in association with past farming methods that gives them their particular symbolic significance. The horse's role has shifted from that of a working animal to an animal used for leisure pursuits which reflect and maintain a link with a past way of life.[32]

With modern agricultural practices under attack for polluting the environment, and modern tractors criticized for compacting the land, some devotees of the working horse not only see it as having a symbolic linkage with the past, but take the view that its resurrection on the farm is essential for the future of sustainable farming. Murt Fitzgerald from West Cork, who works the family farm with Irish Draught Horses, like his father and grandfather before him, regards the draught horse as the 'power unit of the future' and believes that 'There should ... be at least one horse on every farm to supplement the tractor and do the many jobs that the horse does best ... I think we must get back to a more natural and sustainable type of food production – and the draught horse (in my case the Irish Draught horse) is a fundamental part of that rethink'.[33]

CHAPTER FOURTEEN

Cattle

THERE IS NO evidence for wild oxen in Ireland, and the earliest cattle bones, dating from the Neolithic period, are of domesticated cattle. The ongoing importance of cattle is clear from Irish literature from the earliest times, and cattle were at the core of Irish farming throughout the last three centuries.

TYPES AND BREEDS

The notion of a breed with standardized characteristics and a documented pedigree is relatively recent. For most of the eighteenth century, it is probably less confusing to think of types of cattle, which shared similar features in a general way, rather than breeds, but the terms were used interchangeably in contemporary writing. In 1835, Sir William Wilde claimed that Ireland had four native breeds of cattle; the Longhorn, the Kerry, the 'Old Irish Cow' and the Irish Moil.[1]

The Kerry

The Kerry is sometimes referred to as the 'poor man's or the Irish cottar's cow'[2] because of its ability to thrive in harsh conditions (plate 13). Found mainly in the mountainous areas of south west Ireland, the Kerry was recognized as distinctive by the English agriculturalist Arthur Young in the 1770s, and attempts to describe its characteristics survive from 1800.[3] One of the more detailed descriptions in the first half of the nineteenth century was given by David Low, professor of Agriculture at Edinburgh University. In 1845 Low noted:

> Kerry cattle are generally black, with a white ridge along the spine ...
> They have often also a white streak upon the belly, but they are of
> various colours, as black, brown, and mixed black and white, or black
> and brown. Their horns are fine long and turned upwards at the points.
> Their skins are soft and unctuous, and of a fine orange tone, which is
> visible about the eyes, ears and muzzle. Their eyes are lively and bright,
> and although their size is diminutive their shape is good.[4]

Along with its hardy constitution and capacity to subsist on 'scanty fare' the most important quality of the Kerry, for Low, was its value as a dairy cow. Extolling its virtues as a milch cow he maintained:

> In milking properties the Kerry cow, taking size into account, is equal or superior to any in the British Isles. It is the large quantity of milk yielded by an animal so small, which renders the Kerry Cow so generally valued by the cottagers and smaller tenants of Ireland.[5]

However, Low was critical of the apparent lack of a 'progressive' management and breeding system to develop the breed and preserve its purity.[6] Ironically officialdom had taken a tentative step towards recognizing the Kerry as a distinctive breed the previous year, with its first appearance at the Spring Show of the Royal Dublin Society in 1844[7]. Dr Leo Curran identifies this as the beginning of an upsurge of interest in Kerry cattle during the 1840s.[8] One entry, from the show of 1847, illustrates this, even at the height of the Famine.

> Section 4: Kerry
> Rev. Robert Healy, Clongowes Wood College, Co. Kildare [Kerry Bull, King Dan, by Teague a Botha, own brother to Foigh-a-ballogh; dam Noreen, own sister to Paddy Blake, calved on 1st January 1846, reared (without milk) for first three months on boiled potatoes; out on a coarse pasture till within the past two months, when he got a little hay and turnip-tops].[9]

By the 1840s records of breeding were also being kept, although a national register and herd book was not established until the latter end of the century. Tests were organized to compare Kerry cattle with other breeds. Milk production, butter content and amount of fodder required were all tested. For example, at Ballydesmond County Cork, in 1841, Kerry cattle were compared to Galloway and Ayrshire cattle. It was found that the Kerry cattle tested gave 7.5 quarts of milk a day, mid-way between the amount produced by the other cattle, but a pound of butter could be produced from this milk. Significantly more milk was required from the other breeds tested, to produce the same amount, and the Kerry also required less hay as fodder.[10] Interest in Kerry cattle grew during the 1850s because of this relatively high milk yield and butter fat content, and Kerry cattle were exhibited at shows in Paris in 1856 and 1879. By the later nineteenth century, the Kerry's reputation for hardiness was well established. In 1880, one writer described the system of rearing the cattle.

> These little creatures are reared chiefly by small landholders, and at a very early age are turned, without house or shelter of any kind, upon the mountains of their native county … where they shift for themselves as best they can until they are about two years old. They are then collected, sorted, and sold to dealers, who again dispose of them to the farmers in Connaught and other arable districts, who eventually stall-feed them on turnips, hay, oilcake, etc., making them at four years old, weigh from 5–6cwt. of prime beef.[11]

Despite the generalized descriptions such as Low's above, there was also an awareness amongst observers of two distinct types within Kerry cattle. In 1845 it was remarked, 'when an individual of a Kerry drove appears remarkably round and shortlegged it is common for the country people to call it a Dexter'[12] The origin of the Dexter and its relationship to the Kerry has been the subject of ongoing debate. One school of thought claims that a Mr Dexter of County Tipperary, who was reputed for breeding small cattle, started the breed, while another has used archaeological evidence to argue that cattle of the Dexter type have a very long history in Ireland. However, most animal breeding experts believe that the Dexter was an off-shoot of the Kerry.[13]

It was not until the late nineteenth century that formal attempts were made to separate Kerry and Dexter cattle into related but distinctive breeds. By 1872, the Kerry Agricultural Society was said to distinguish between the breeds in its show classes, but it was noted that governing bodies such as the Royal Agricultural Society of Ireland and the Royal Dublin Society did not.[14] The Royal Dublin Society had introduced separate classes for Kerry and 'Dexter-Kerry' cattle in its 1863 Spring Show but this appears to have been abandoned in subsequent years and it was not until 1876 that a separate class for Dexters was finally established.[15] The official seperation of the Kerry and Dexter into distinct breeds gathered momentum during the remaining years of the century. In 1887, the Royal Agricultural Society of Ireland started a register of Kerry and Dexter cattle, and in 1890, a Kerry and Dexter Herd book was established – 1,297 animals were registered: 118 Kerry bulls, 942 Kerry cows, and 137 Dexters.[16] Colours for the breeds were fixed and conditions for entry into the

108 Dexter cow with the bull 'Big Bill Campbell'.

Herd-Book stipulated that Kerry bulls should 'be pure black with the exception of a few grey hairs about the organs of generation in animals of exceptional merit'. while Kerry cows and heifers should be 'pure black with the exception of a small portion of white on the udder in animals of exceptional merit' (plate 14). Dexter bulls and cows could be either black or red, with a little white (fig. 108).[17]

A physical description of the two breeds contrasted them as follows:

> The *Kerry* cow is elegant and deerlike, with light limbs and body, not fleshy, light at the shoulder and deeper in the hind quarters; her skin is soft and unctuous; her head is light and graceful, with bright eyes and ears; her horns are white with black tips, planted widely, not thick at the base, and rising outwards and upwards, often turning inwards towards the points.
>
> The *Dexter* cow has very short limbs and a deeper body; she is therefore less elegant but no less pleasing; her head is stronger, but very clear cut, shorter below the eyes and broader in the muzzle; her horns are thicker, and usually, after rising upwards, bend backwards towards the points. She is more fleshy than the Kerry, but, even so, looks a better milker. Her udder is larger, and extends farther forward.

> The *Kerry* and *Dexter bulls* differ in much the same way as the cows –
> the Kerry being light in limb and body, and the Dexter short-legged and
> stout: the one of the dairy and the other of the beef type; the horns are
> thinner and somewhat upright in the Kerry, and thicker and a little more
> horizontal in the Dexter.[18]

Both the Kerry and Dexter were seen as 'excellent milkers' and 'ready
fatteners'.[19] However the Dexter was particularly recognized as possessing
excellent beef producing qualities 'fattening rapidly on even middling pasture'
and producing beef which was 'exceedingly, fine and well flavoured.'[20]

The Kerry breed became more popular during the latter part of the
nineteenth century when leading families in Ireland began to champion the
breed. Keeping a pedigree herd increased social status, particularly when
Anglo-Irish families started to keep them, making it fashionable to be associated with the exhibition and breeding of the native cattle.[21] The association
of leading families with the breeding of pedigree cattle is reflected by the Kerry
milk records in which the list of owners included John Hilliard, Killarney;
Arthur R. Vincent, Muckross Abbey, Killarney; the knight of Kerry, Valencia
Island; the countess of Clanwilliam Montalto, Ballynahinch, Co. Down; John
Neill, Aghadoe Cottage, Killarney; the Hon. Mrs Vandeleur, Cahiracon,
Ennis, Mrs E. Robertson, Limavady; Stephen E. Brown, Ard Caein, Naas; and
the Duke of Leinster, Carton, Maynooth.[22]

The popularity of the Kerry and Dexter spread beyond the shores of
Ireland. A major force in promoting the cattle in England was James
Robertson of La Mancha, Malahide, County Dublin. He brought them to
prominence by exhibiting good specimens of his own herd at the Royal
Agricultural and the London Dairy Shows. Robertson also developed a
thriving export trade to England from 'carefully selected young and unde-
veloped cattle of the purest and best sorts to be found at the Killarney and
other Western Fairs'.[23] As with the Kerry and Dexter breeds in Ireland, the
cattle were championed by leading people in England. They established a
reputation not only as 'showyard celebrites' and 'general-purpose animals
(suitable for both dairy and beef production) of high merit' but also as
'ornamental cattle' graceful enough to adorn the large estates of England.
Dexters were particularly popular as ornamental cattle (fig. 108) and were said
to be 'especially admired by ladies'. Among the better-known herds of Dexters
kept by leading families was that owned by King Edward. An English Kerry

and Dexter Cattle Society was founded in 1892 and the first volume of the Herd Book published in 1900. As in Ireland, the emphasis was to maintain Kerry and Dexter breeds distinct with pure pedigrees. Regulations in the Herd Book stipulated that 'a cross between the Kerry and the Dexter is considered a half-breed, and cannot be entered'.[24] The 1890s also saw the export of both breeds further afield with registered herds being established in Scotland, South Africa, America and Australia.[25]

Unfortunately, the desire to have pure lines of breeding was to have far-reaching consequences for the survival of Dexter cattle in Ireland. On 14 July 1917, the Kerry and Dexter Cattle Society of Ireland was formed in Killarney, aiming to 'promote the interests of Kerry and Dexter cattle breeding in Ireland'.[26] Less than two years later at the Annual General Meeting of the Society on 11 June 1919 in the Board Room of the Royal Dublin Society it was decided to change the name to The Kerry Society of Ireland. The reasons given for the alteration were:

> Herds of pedigree Dexter cattle have practically ceased to exist due to the difficulty of breeding them pure … when Dexter cows are mated with a Dexter bull a large proportion of the progeny are either still born or deformed-monstrosities … as far as Irish breeders are concerned their whole attention is now directed to the development of the Kerry breed.[27]

The small size of the Dexter and the genetic tendency to produce a monstrosity or 'bull-dog calf' was related to a condition known as *achondroplasia*, or 'dwarfism'.[28] The declaration of intent, to develop the Kerry by the newly founded society sounded the death knell for the pedigree Dexter in Ireland and its eventual extinction. It was not until the 1980s and its reintroduction into Northern Ireland by members of the Rare Breeds Survival Trust N.I. Support Group, that the Dexter gained a foothold back in its native island, where its ability to produce prime beef has found a niche in the rare breeds meat market. However, the problems related to the genetic tendency to produce 'bull-dog' calves did not affect the breed's popularity in England. By 1890, the Dexter was gaining in popularity over the Kerry. Dexter cattle were hailed as 'miniature dairies' and Dexter milk was claimed to be richer than Kerry milk.[29]

During the early 1900s, enthusiasm for both breeds began to decline in England. Undoubtedly the decision by the Kerry Society of Ireland to focus

109 Irish Moiled cattle in a byre at the Ulster Folk and Transport Museum.

on the development of the Kerry breed was a major factor in the demise of the Dexter in Ireland. However Leo Curran suggests that underlying economic and political factors also played a crucial role in the loss of interest in both breeds. He argues that Kerry farmers did not have the resources to develop the finer points of the breeds, while English breeders lost interest in Irish cattle, partly because Ireland was moving towards independence from Britain and political change meant less patronage from the landed families who had previously formed a core of breeders.[30] Despite these setbacks, by 1931 there were still 45 herds of Kerry cattle in Ireland, fifteen of them in the Kerry/West Cork area. However, survival of the breed reached a crisis point in the 1980s, when numbers dropped to 28 registered bulls and 143 cows. An ongoing conservation strategy was developed between the Royal Dublin Society, the Kerry Society of Ireland and government, which included premium schemes for private owners, the establishment of herds of Kerry cattle in a number of national parks and the promotion of the Kerry as part of the national heritage.[31] The conservation of Dexter cattle was almost all carried on in England and South Africa until its reintroduction in the 1980s mentioned above. Hopefully the survival of the Kerry and Dexter is now assured.

The Irish Moil

The Irish Moil is a dual-purpose breed suitable for both beef and milk production (fig. 109). The name comes from the Irish word *maol*, meaning 'hornless'. Moiled, or hornless types of cattle have a very long history in Ireland. Hornless skulls have been found at Early Christian sites, such as Lagore Crannog, near Dunshaughlin, County Meath. The number of such skulls is more than would occur naturally, and may have been the result of selective breeding.[32] The Moil was recognized as distinctive in 1835, when their colours were described as dun, black, white, or coloured. In 1845, the Scottish agriculturalist, Professor D. Low also mentioned Moiled cattle as 'the Polled Irish breed', which was scarcely known in England. He described them as light brown in colour, and said that they were most common around the Shannon. He saw the Moil as having great potential. 'Had attention been directed at an earlier period to its preservation, Ireland might now have possessed a true Dairy Breed, not surpassed by any in the kingdom.'[33]

During the nineteenth century, attempts were made to develop the Moil as a breed, by a few individuals. Dr P.H. Fox of Benown, Athlone, for example, bred Moils during the later nineteenth century. He kept a herd of the cattle for nearly twenty-five years. He described the characteristics of Moils that he thought distinctive. 'If the crown of the head projects upwards and is finely drawn almost to a point on the top of the head the breed is decidedly good … the chin projects a little, the lips are thick and heavy, and the nose is somewhat short … The colour is red, the nostrils are usually brindled and the tips of the ears are brown.' Fox praised the Moil for its milking qualities, which included a high butterfat content.[34]

During the early years of the twentieth century, the Moil seems to have been most common in the north of Ireland, particularly Tyrone, Armagh and Sligo. A Major John Murray of Ballyronan Park, Maynooth, County Kildare, was said to have used cows from these counties when developing a herd. The colours noted were pepper and salt, tending to black roan. Captain Herbert Dixon was also named as a noted breeder.[35]

Attempts to create a standardized breed of Maol or Moiled cattle, continued in the early twentieth century. The following appeared in the *Irish Homestead* on 12 April, 1919.

Society for the Preservation of the Maol or Hornless Breed of Cattle

A meeting of this society was held on the 23rd of last January and again on the 13th ult … It was decided to offer a prize at the Drogheda Show next September … The winner must, of course, be hornless and have a long head and a high and pointed crest. The colours favoured are dun, yellowish red and roan, and the colours barred are those characterizing the Angus, Galloway, Hereford and Red Poll.

From this it is presumed ordinary Shorthorn colours will not be excluded if an animal is otherwise unobjectionable. [The report congratulated the members on the discovery of many more Maolines and of their more general distribution throughout the country than was anticipated].[36]

However, hopes that the Department of Agriculture and the Royal Dublin Society might take an active interest in the breed came to nothing.

An Irish Moiled Cattle Society was successfully established in Belfast in January 1926, after a meeting in the Old Town Hall, with Captain Herbert Dixon as President, and Captain J. Gregg as Registrar and Secretary. The aim of the society was to establish a hardy dual-purpose animal, suitable for the hill farms of Ulster. It would be able to survive on 'indifferent' pasture, give a reasonable supply of milk, and when crossed with beef cattle, be capable of producing good saleable stores. The society established a herd book and the first animal to be registered was Donegal Biddy, 23 years old and owned by John O'Neill, Crumlin, County Antrim. In its early years the society accepted cattle of various colours for registration. However, in 1929 it was decided to aim for red or roan with a white stripe down the back as the official colouring. This often occurred in the Moil, and was described by R. Wallace as a common characteristic of Irish cattle. It was a sufficiently striking characteristic to be identified by an Irish word *droimeann*, meaning 'whitebacked' (plate 15).[37] Another characteristic colouring which has survived in some Moils, is of equal interest for folklorists. Some Moiled Cattle are almost white, with red muzzles and red ears. In early Irish literature, such cattle were described as magical.[38]

A bull premium scheme was established and good quality Moiled bulls sent out to various districts and made available to local farmers and cattle exhibited at Belfast's Royal Ulster Agricultural Show. With the death of Captain Gregg, the main driving force behind the society, and the onset of the Second World

110 Irish Longhorn cattle, from Library of Useful Knowledge, *Cattle* (London, 1834).

War, the Irish Moiled Cattle Society went into decline. Mr A.E.H. Gillespie got the society back on track in 1948, and this was boosted the following year when Captain Percival Maxwell of Downpatrick founded the Ballydugan herd.[39] Much debate occurred amongst Irish Moiled breeders as to whether the Moil was brought to Ireland by the Vikings or taken by the Vikings from Ireland to Scandinavia. Captain Gregg was said to be firmly of the opinion that present-day Finnish polled cattle were descendants of Irish Mois.[40] In 1950 it was decided to introduce the Finnish bull 'Hakku' to the Ballydugan herd and and he was claimed to have produced some very good calves. The society went into decline again in the late 1950s, mainly due to the requirement of the Agricultural Act (N.I.) 1949 that only bulls which could be licensed were those from a cow with a recorded high milk yield. Unfortunately most of the Moil breeders do not seem to have kept milk records and as a consequence the breed declined.[41] By the 1980s only two herds of Moils survived. These were owned by David Swann, Dunsilly, County Antrim, and James Nelson, Killyleagh County Down. In 1982 the Society was revived by a few dedicated owners of Irish Moils, the Belfast Zoo, and the National Trust, with financial help from the Rare Breeds Survival Trust.[42] A new herd book was started in

1983 and in the following years institutions such as the Ulster Folk and Transport Museum became involved in the conservation of the breed. The museum purchased cattle for its living farm project, and in association with the Rare Breeds Survival Trust Support Group staged the annual Rare Breeds Sale and Show at which Irish Moiled Cattle were sold. By the 1990s the breed had been rescued to the extent that Moiled cattle were being exported and were fetching some of the highest prices at rare breeds sales in Ulster and England. Like Dexter cattle their meat also found a niche in the rare breeds meat market.

Irish Longhorns

An Irish type of Longhorn cow was identified in the early nineteenth century, particularly in the western midlands (fig. 110). Sir William Wilde outlined their characteristics.

> The Irish longhorns – resembling the Lancashire and the Craven; in some cases the horns were wide-spreading and only slightly curved, but frequently the horns were so completely curved inwards as to cross in front of or behind the mouth, or pressed inward so much towards the cheek as to become a source of great irritation to the animal, and to require amputation; they were generally a red or brindled colour, had large bones, grew to a great size, particularly as bullocks, and their drooping horns, sloping gracefully under the skin, gave them a particularly calm expression of face: they were covered with a plentiful supply of hair … [They were] an exceedingly hardy race of cattle, never requiring fodder except when the ground was covered with snow … not much used as milkers; they were the principal cattle sent to the Dublin market or exported thirty years ago; their hides were of great value … they might justly be termed 'the Connaught ox'.[43]

There has been ongoing, sporadic discussion about the relationship between Irish and English Longhorns. David Low thought that the English breed of Longhorns originally came from Ireland, and the English improver Robert Bakewell may have used some of these in developing his famous breed. This was also the opinion of the *Standard Cyclopedia of Modern Agriculture*, which suggested that, 'the old-time breeders [were of] … the opinion that during transit to our metropolitan markets many of their best specimens were arrested by English farmers, who thus laid the foundations of their herds with Irish-bred stock'.[44] It is clear though that, at least from the late eighteenth century, breeders wishing to improve Irish Longhorns, brought over cattle from Britain. In 1794, for example, it was claimed that

> The Irish cattle [are] ... a mixed breed between the longhorns and the Welsh or Scotch, but most inclined to the longhorns, though of less weight than those in England ... Irish breeders have brought long-horned bulls and heifers at very high prices from Lancashire, Leicestershire, Warwickshire etc., particularly the Mr Frenches and other spirited dealers from Roscommon and different parts of the West of Ireland which have been of great advantage in improving their breed.[45]

In the early nineteenth century, it was claimed that dual-purpose Longhorn cattle predominated in some areas, such as the Burren in County Clare.[46] By the mid-nineteenth century, however, the Longhorn breed was claimed to be nearly extinct in Ireland. In 1872, one observer claimed that 'The last time I saw Longhorns exhibited in public was at the Spring Show of the Royal Dublin Society, in 1858, when Lord de Freyne exhibited a pair of two-year-old oxen in the fat stock classes; and so little was the breed known then that I recollect some persons present enquiring what "foreign breed" Lord de Freyne's bullocks belonged to.'[47]

One of the last big herds of Longhorns in Ireland was kept by the earl of Westmeath on his estate in County Galway where it was claimed 'to have been bred there in a pure state for over two hundred years.[48] Describing Lord Westmeath's herd in 1910 *the Standard Encyclopedia of Agriculture* noted:

> Lord Westmeath's herd in County Galway, Ireland, is still in a very flourishing condition, though it is but little heard of [in Britain] ... As late as the year 1902, however, this herd furnished steers good enough to carry off second prize at the Royal Dublin Society's Christmas Show in a very strong mixed class of Angus, Herefords, and Shorthorns, as well as numerous crosses.[49]

The Old Irish Cow and Other Breeds

The 'Old Irish cow' is the least documented of the breeds identified by Wilde. It had a range of characteristics, which have been well summarised by Leo Curran. Wilde's own description is as follows:

> Old Irish Cow – of small stature, long in the back, with moderate sized, wide-spreading slightly elevated and projecting horns; they could scarcely be called long-horned, and they are certainly not short-horned; this breed was of all colours, but principally black and red. They were famous milkers, easily fed, extraordinarily gentle, requiring little care, and were, in truth, the poor man's cow … but they did not fatten easily, and when beyond a certain age seldom put on flesh: they abounded in all parts of the plain country.[50]

Unfortunately, there is very little else recorded about these cattle. On smallholdings, nineteenth-century observers often pointed out that nothing had been done to improve the 'poor stunted' animals kept locally.[51] Since the eighteenth century, the cattle industry in Ireland has been dominated by introduced breeds. The catalogue for the Royal Dublin Society's Spring Show in 1831 lists classes for Leicester, Durham, Hereford, Ayrshire, and Devon cattle, with mention of the Alderney (Channel island) breed.[52] Between 1850 and 1900 there was a 60% increase in cattle numbers in Ireland, mostly of imported breeds such as Shorthorns, which by 1900 had become the most common breed throughout Northern Europe. Ireland was known as a centre of excellence for Shorthorns. The Department of Agriculture and Technical Instruction for Ireland established a National Shorthorn herd, and the breed remained dominant until the 1930s, although encouragement was also given for the purchase of other breeds, such as Herefords, Aberdeen Angus, and Galloways. Friesian cattle became common in dairying areas during the 1950s, while in more recent decades breeds such as the Charolais and the Polled Hereford have reflected a move towards greater diversity of breeds, mostly imported from mainland Europe.[53]

CARE AND MANAGEMENT OF CATTLE

Kevin Whelan has distinguished two main regions of cattle farming. Both developed on the well drained carboniferous limestone areas of the midlands and west of Ireland. North-East Leinster and East Connacht formed the core of one of these areas, but it also extended into Offaly, Longford and Westmeath. This was a region of good land and large farms. By the late eighteenth century, its main business was the fattening of cattle for the Dublin and English markets. The big graziers bought in young cattle from the west in October, and sold them on the following summer.

The second area for large-scale cattle farming specialized in dairying and butter production. It stretched over an area of good land from Kilkenny to West Limerick and North Kerry. Farms were somewhat smaller than those in the Midlands dedicated to cattle fattening, and the more labour intensive work involved meant that there were more workers employed.[54] Herds of over 1,000 cattle were recorded in rich grazing areas, such as the herds of Mr McCarthy of Springhouse in County Tipperary, mentioned earlier. This regional specialization remained, and indeed intensified during the twentieth century.

At the other end of the farming scale, the ownership of a cow was also crucial, providing both manure, essential for the successful cultivation of potatoes, and milk, which along with potatoes could provide a healthy if monotonous diet. It was in these areas that the seasonal movement of cattle to hill pastures, or booleying, was common. Concern with care of cattle on small holdings was shown by articles in the *Dublin Penny Journal*, such as 'Method of Keeping a Cow upon a Quarter of an Acre'.[55]

The movement of cattle to hill pastures during the summer was often associated with the rundale system in the late eighteenth and nineteenth centuries. These large-scale movements declined with the spread of enclosure, but in one area, the Burren of County Clare, unique environmental conditions led to the practice of 'wintering out' cattle on the hills, which is still practised.

> The limestone of the Burren hills is said to act like a giant storage heater, building up heat in summer and dissipating it in the winter months when the surrounding air is cooler. As a result, a relatively warm, dry lie is ensured for outwintering livestock, with the thin sparse soils scarcely masking the limestone's warmth, while proving very resistant to water-logging, muddying and erosion. Though nutritionally somewhat limited, the calcium rich vegetation, does promote strong bone growth, while the range of minerals provided by the diverse flora promotes animal health.[56]
>
> The weakness of cattle after being kept indoors during winter has been recorded in several countries. In Ireland, milk cattle kept indoors during winter had to be carried out to the fields in spring. In the Mourne Mountains in County Down, these cattle were sometimes known as 'lifters.'[57]

Best practice for stall feeding was outlined by Thomas Baldwin in 1877. For the first ten days, soft turnips were recommended, then more fattening root

111 Buttermaking, County Antrim, *c.*1910 (UFTM L832/10).

crops, the roots being sliced first to prevent choking. Straw and hay, with oil-cake or corn were also to be fed, the oil-cake broken very fine before feeding. The oil-cake could be made from linseed, rape, or cotton.[58]

At the level of small-scale farming, the rearing of young calves was often seen as women's work. As late as 1947, the importance of this was emphasized, as well as the extent to which it was undervalued. John Mogey commented, 'Should these women ever combine to demand an economic return for their work, the whole livestock industry of Ireland would have to be re-organised.'[59]

BEEF AND DAIRY PRODUCE

The Irish beef trade reached a peak of 144,597 barrels in 1815, but tended to decline after this. However, steam navigation, established in 1815, meant that the export of live cattle to Britain was made much easier, and this more than made up for the decline in the export of salted beef.[60] Exports have continued ever since, despite duties and quotas imposed by Britain during the 'Economic War' of the 1930s. One result of the latter was the development of large-scale smuggling of cattle across the Border between the Irish Free State and Northern Ireland. It was estimated that in 1935, around 100,000 cattle were exported from the south to the north in this way.[61]

112 A roadside butter market, County Cork, 1890s, from
Pictures of Irish life (Cork, 1902).

Before the establishment of creameries at the end of the nineteenth century, butter was produced on the farm, mostly by farm women (fig. 111). The biggest regional difference in making butter was between the north and south of the country. In the North, the whole milk was churned, while in the South only the cream was used.[62]

In 1877, Thomas Baldwin described the best methods of making butter. The most important dairy utensil was the 'cooler'. This could either be a shallow staved oak vessel, or earthenware – Baldwin advised Staffordshire pottery, rather than Scotch ware glaze, as the latter tended to peel in hot water. Where cream was used, he advised that the milk should be strained and put in a cooler for between a day and a day-and-a-half. It was then removed with a skimmer, and poured into deep 'cream jar' where it was allowed to ripen for about three days. If the whole milk was being churned, it stayed in the cooler for between 12 hours and one day. Two or three milkings were then put into a large vessel until a stiff *brat* appeared on the surface. This usually took about a day-and-a-half in summer.

The most common churns on small farms in Ireland were plunge operated. Baldwin agreed that the plunge churn produced excellent butter, but pointed out that it required a lot of time and labour, especially when whole milk was churned. By 1877, he claimed that most dairy farmers were using barrel churns, which produced butter much quicker. He estimated that 12 quarts of milk should produce 1 lb of butter.[63]

Butter-making on the farm was women's work, and women were able to barter or sell the butter they produced at home (fig. 112). However, the large-scale export of butter to countries such as the West Indies, Brazil, Spain, etc., was well established by the mid-eighteenth century, largely, it was claimed, because the art of salting butter was better known in Ireland than in any other country.[64]

By the end of the eighteenth century, butter was Ireland's largest agricultural export. The biggest centre for the butter trade was the Cork Butter Exchange, founded in 1770. The exchange was established to regulate the butter trade. The exchange developed a global trade to the extent that it was claimed to be the largest centre of its kind in the world.[65] The system of checking butter in Cork was described in 1852.

> Every firkin of butter that passes through the Cork Weigh-Houses – and nearly every firkin that enters this city passes through it – is rigidly examined and its quality determined, and when this butter is received by the foreign buyer he has a sufficient guarantee as to the character and quality of the article in the well-known brand upon its cask. [Butter is graded between first and sixth quality. In this way the] farmer, the merchant, and the foreign buyer are equally protected against fraud.[66]

The port of Waterford was also a major centre for butter exports from towns such as Tipperary and Clonmel.[67] The relative importance of butter declined after 1820, but the number of dairy cattle remained at about 1.5 million until the early twentieth century. The market for butter remained buoyant until the 1870s, when other countries began to compete effectively for the British market. The co-operative movement (the Irish Agricultural Organisation Society) began to establish creameries throughout Ireland from the 1890s onwards. By 1900, there were 236 Dairy Societies, with a membership of 26,577. Farmers supplying milk to the societies were said to have increased their profits from dairying by between 30 and 35%.[68]

The co-operative's ideal of creating a 'rural civilisation' meant that some early attempts were made to extend the benefits of the co-operative creameries to everyone. R.A. Anderson, one of the movement's leaders, described one initiative.

A very pleasing feature in the development of the creamery system is the opportunity it has given to labourers to become cow owners. Numbers of them now have cows, and one case has been reported where a man living in an ordinary way-side cottage with one acre of land, has been enabled to own eight milch cows, from the milk of which he has realised £70 in cash during the last year. From grazing one cow by the roadside – on the 'long farm', as it is called in the country – he was enabled to buy additional cows and rent grazing for them through the profits he derived from the Creamery.[69]

The Department of Agriculture and Technical Instruction also took effective measures for ensuring high quality and efficient marketing of Irish butter. In the first decade of the twentieth century, the Department provided loans to creameries for the purchase of pasteurizing plants, and also started regular inspections of creameries. The 1906 Butter and Margarine Act provided for the registration of butter factories. The Department also organized classes in dairy technology and creamery management, and along with the co-operative movement, improved the reputation of Irish butter in Britain and elsewhere.[70]

By 1914 there were around 350 dairy or creamery societies, and creameries remained the most successful part of the co-op movement.[71] However, only 39% of Irish butter was produced in creameries. Most was still produced on the farm.[72] The dairy trade with Britain flourished during the First World War, but slumped afterwards. During the War of Independence, and especially in 1920, British forces burnt creameries as a reprisal for IRA attacks. By the beginning of September 1919, nineteen creameries and separating plants in counties Tipperary, Limerick and Clare had been damaged or destroyed.[73] Creameries in Border areas suffered during the 1930s. International competition held back the development of the dairy industry, until Ireland's entry into the EEC.

Cheese-making was very developed in medieval Ireland, but references to cheese making become less common in the modern period. However, it seems to have continued on some bigger farms. In the 1830s, for example, it was claimed that 'Carrickfergus was formerly celebrated for its cheese which to the present day bears a respectable character and is held in considerable estimation. New milk cheese is made in large quantities but not so much as formerly.'[74] By the 1870s, it was claimed that cheesemaking had disappeared from Ireland. Thomas Baldwin thought that this was bcause farmers preferred the ready sales from milk and butter, to cheese, 'which requires months before it is brought to market'.[75]

CHAPTER FIFTEEN

Sheep

D URING THE LAST three centuries, sheep farming in Ireland has changed from relatively small scale production, to a highly organized part of farming. In the eighteenth century, most sheep had not been improved by systematic breeding, and observers were generally unimpressed by them. In 1807, the agriculturalist George Cully was dismissive.

> The Irish Sheep, a pretty large sample of which I saw at the great fair of Ballinasloe … I am sorry to say, I never saw such ill-formed ugly sheep as these; … [they] are supported by very long, thick, crooked grey legs; their heads are long and ugly, with large flagging ears, grey faces and eyes sunk; necks long, and set on below the shoulders, breasts narrow and short, hollow before and behind the shoulders; flat sided, with high narrow herring-backs; hind quarters drooping, and tail set low. In short, they are in almost every respect contrary to what I apprehend a well-formed sheep to be.[1]

As with other livestock, attempts to develop sheep breeds with standardized characteristics only began seriously during the later eighteenth century. Arthur Young identified a 'long-legged Tipperary' as a distinct type, but in 1780, he also noted that Lord Montalt of Dundrum, County Tipperary, had begun to change his own sheep by introducing several rams of the shorter legged Leicestershire breed.[2] Leicester sheep, along with Scottish Blackface, were the most common introduced breeds farmed before the 1850s. However, by the mid-nineteenth century, several local types of Irish sheep were recognized as worth improving. These were of two main varieties: small, hardy mountain animals, such as the Kerry Hill type, or larger lowland animals, notably the Roscommon sheep.

ROSCOMMON SHEEP

The Roscommon was the most successful Irish breed of sheep (fig. 113). Agriculturalists recognized it as distinctive from the early nineteenth century, when it was agreed to be the product of cross-breeding between native Irish sheep and an improved English breed, the Leicestershire mentioned above.

113 Roscommon sheep (UFTM L882/10).

During the eighteenth century, the English agricultural improver, Robert Bakewell, had begun to improve sheep on his farm at Dishley Grange in Leicestershire. From the middle of the eighteenth century, large numbers of Dishley Leicesters, also known as New or Improved Leicesters, were imported into Ireland. Bakewell himself actively promoted this trade. Lord Roscommon, later the earl of Sheffield, was a close friend of Bakewell, and was influential in establishing the Dishley breed in Ireland. The earl had large estates in the west of Ireland, where he encouraged his tenants to use Bakewell rams. The effects of this initiative were rapid. Hely Dutton wrote in 1824,

> When I first came to Ballinasloe, having heard so much of Connaught sheep, I was not a little surprised at seeing such multitudes with thick legs, booted with coarse wool down to their heels, and such a bushy wig of coarse wool on their heads, that you could scarcely perceive their eyes; at present they have nearly all disappeared, and given place to a fine breed, not to be equaled by the general stock of long-wooled sheep in England.[3]

By the 1830s, it was claimed that the use of Dishley and Leicester sheep, along with careful selection of local rams for breeding, mean that 'no one would recognise a single trace of the uncouth, worthless animal which Cully depicted.'[4]

The new sheep were recognized as a distinct breed. A Roscommon Sheep Breeders Association was established in 1895, and systematic registration and marking of the sheep began. In 1902, the Roscommon was described as 'the only native breed of sheep that Ireland can claim'.[5] In 1908, the sheep were described as large and hornless with a long-wool, bright lustrous fleece. The head was well-shaped, with a long white face, fine, medium length ears and sometimes a tuft of wool on the forehead. The tail was well-hung and the legs broad and strong. Roscommons could be kept on high, sparse ground, but those on the lowlands were described as 'big, useful sheep, though lacking in symmetry.' In the 1870s, the breed had been improved further by a 'moderate infusion of the blood of English long-wooled breeds, notably the Leicester'. In the early twentieth century, Roscommons were described as hardy, excellent grazers. They were slower to mature than other long-wooled breeds, but their mutton was said to be better. If left to forage, the fleece would yield eight to twelve pounds of wool. If the sheep were well cared for, and partly hand fed, twelve to sixteen pounds might be expected. The staple was typically about seven inches long.[6]

Despite their reputation, however, the 1924 Flock Book of the Roscommon Sheep-Breeders' Association showed that the old society was in serious trouble, with only five breeders registering rams. In that year, 187 unregistered ewes were admitted by inspection, in what proved to be a vain attempt to increase the registered population.

GALWAY SHEEP

In the early part of the twentieth century, Galway sheep were seen as slightly smaller, neater, and more easily fed variant of the Roscommon breed (fig. 114). The breeds are closely related. Like Roscommon sheep, the Galway shows a resemblance to illustrations of Bakewell's sheep, and it appears closer to the Dishley breed than many the English breeds which were influenced by similar programmes of cross-breeding. However, the interrelationship between Galway and Roscommon breeds is not accepted by everyone, and enthusiasts claim the Galway as a distinct breed.[7] The separation of the two breeds may have arisen because of local rivalries. Most of the fifty members of the Roscommon Sheep society listed in the register of 1924 were from County Roscommon, with no breeder listed from Galway. The Galway Sheep Breeders Society was founded by a group of breeders in Athenry, County Galway, in 1923. At that time over 6,000 ewes and 200 rams were inspected and the best 10% were admitted to

114 Galway sheep in the Ulster Folk and Transport Museum.

the new flock book. This number of sheep clearly did not suddenly appear. The new society seems to have selected good examples of the smaller type from the existing population of 'Roscommon' sheep.

The Galway Sheep Breeders Society continued to admit flocks after inspection, provided that they were true to type, of the required quality and in good health, until 1953, when the flock book was closed. To ensure quality, the Society introduced inspection of all registered lambs, tattooing of lambs by the Society recorder and, to ensure accuracy of pedigrees, a ruling that no Galway breeder could keep a ram of another breed on his premises. Numbers of Galway sheep were never huge, but the breed still survives, and registration continues in Athenry. Galway sheep are now recognized as a rare breed in Britain, and there is now a Galway Sheep Breeders Society of Great Britain.[8]

Roscommon and Galway sheep were not the only varieties claimed to be native to the plains of East Connacht. Some enthusiasts claimed 'Ballinasloe' sheep to be distinctive. They were said to be common in Galway, Roscommon and Mayo, but there is no clear account of the points that might distinguish them from Roscommon or Galway breeds.[9]

Several types, or 'breeds', of mountain sheep were identified during the last three centuries. Arthur Young, for example, noted a small breed of sheep in the Galtee mountains 'which are as delicate when properly fattened as the welch (sic), and of so hardy a breed as to live upon heath, furze, etc., in winter as well as summer'.[10]

WICKLOW SHEEP

Two distinctive types of sheep were recorded in County Wicklow, both of which have now disappeared. In 1845, David Low praised the mountain sheep of Wicklow, for the quality of their wool and mutton (plate 16). He thought the 'pure Wicklow mountain breed' deserved to be preserved, but foresaw that cross-breeding with breeds such as the South Down and Cheviot, would mean that in a few generations the 'indigenous race' would cease to exist.[11] This proved to be the case. Cross-breeding produced Wicklow-Cheviot sheep, which were claimed to be the best mountain breed of sheep kept in Ireland.[12] They were said to be descended from native fine-wooled native sheep, crossed with Cheviot rams imported about 1850, by two Wicklow landlords. They were small, all-white sheep, said to be very active and hardy, with the ability to maintain themselves on sparse mountain grazing through the winter.[13] In 1902, the breed was said to be coming more into favour for breeding. If ewes were brought to a good grazing district, and crossed with a pure-bred ram, particularly of the Shropshire or Oxford Down breeds, the resulting lambs were claimed to be of fine size and prime quality.[14]

In 1922, R.F. Scharff described a type of short-wool sheep, also particularly associated with the Wicklow Mountains. He quoted a source from 1806, which stated that they were generally known as 'cottagh sheep' and that the Bradys of Glenmacnass kept large flocks of the pure-bred stock. The sheep had a small head, narrow face, and short, round and pricked ears. The head and face were smooth and covered with short hair, the wool extending only to the junction of the neck and head. It had a long neck, but the general proportions were good except that it was rather too slender. The legs were small and clean and not very long. The tail had remarkably coarse hair, coarser even than the long-wooled sheep. The fleece was coarse or wavy and occasionally matted, yielding from two to three pounds of wool with a fibre about two inches in length. The wool was made into flannel in the 'flannel hall' in the County Wicklow town of Rathdrum. Scharff gave no source for this description, but says a similar breed had been found in the Connemara Hills in County Galway.[15] These 'mountain sheep of the county Mayo' were described in 1902 as, 'still more of the Cheviot type than those of the county Wicklow; they are smaller in size, longer in the neck, and much less symmetrical, and are slower to fatten.' Defects in the sheep were said to be largely due to bad management, and inbreeding.[16]

KERRY HILL SHEEP

Kerry hill sheep (plate 17) were described in 1845, by David Low.

> The Kerry breed of sheep, notwithstanding of neglect and insufficient food, exceeds in size the breed of Wales, of the Wicklow Mountains, and of many of the Old Forests of England. The horns are generally small and crooked, and sometimes wanting in the female, although some of the allied varieties of other parts have the horns large and spiral. The wool is coarse, and hairy on the haunches, and to a certain degree on the back, but on the sides it is very short and fine. The white colour of the fleece prevails but there is a constant tendency to the development of the darker shades; and the whole Sheep would become black and brown were it not for the choice by breeders of those which are white. These sheep are in a remarkable degree wild and restless in their habits. In shape, eye, neck, position of the head, and general aspect, they approach the Antelope or Deer tribes more than any other Sheep of this country. They fatten so slowly, that, even after they have arrived at maturity of age, they require a long time to be fully fat … however … they are fit for the butcher when their external appearance would indicate that they are still lean. Their mutton is juicy and of good flavour, which causes them to be greatly valued for domestic consumption.[17]

Low argued that the quality of the mutton was a valuable asset, but not enough to justify extending the breed. In 1902, Kerry mountain sheep were dismissed as resembling 'Scotch' sheep, but inferior in quality.[18] It is not clear when they disappeared, but none appear to survive today.

MOURNE SHEEP

In 1902, 'Scotch Hornies' were said to be the dominant sheep in the north of Ireland, in Counties Down, Tyrone, Armagh and Derry. It was claimed that they produced the lowest grade of wool of any sheep in Ireland.[19] The difficulty of deciding when livestock belong to a distinctive breed is well illustrated by the case of the 'Mourne' sheep of County Down. Sheep farmers around the Mountains of Mourne claim that their sheep are different from, and superior to, Scottish Blackface sheep (fig. 115).

In 1802, 'the native breed of sheep of the mountainous areas of County Down' was described as 'very well made in all points, finely woolled and much

115 Mourne ram in the Ulster Folk and Transport Museum.

prized for its mutton'.[20] However, many people saw Mourne sheep simply as a variant of the Scottish Blackface. A correspondent to the *Irish Farmer's Gazette* in 1863, for example, claimed that 'whatever name they are known by, they are derived, for the most part, from the refuse of Scotch flocks, and bear but a faint resemblance to what are known in Scotland as "Linton" sheep'.[21] However, breeders in Mourne insisted that their sheep were distinctive, and superior. A County Down breeder, writing to the *Irish Farmer's Gazette*, under the pen name Curieux of Ponytzpass, asserted that 'As Signarelle says, Il y a des fagots et fagots – there are blackfaces and blackfaces. We have none in Ireland that I know of to equal those of the Mourne Mountains for breeding.'[22] In the twentieth century, the Department of Agriculture for Northern Ireland also recognized Mourne sheep as distinctive, slower to mature and hardier than the Blackface.

The reluctance of experts, and some sheep farmers outside the Mourne area to recognize the sheep as distinctive, is matched by the rather contemptuous attitude of Mourne farmers to the Scottish Blackface. For example, Nora Cooper, a Mourne woman, has listed the differences between a 'Bad Scotch type ram', and a 'Good Mourne type ram,' in terms that allow no doubt as to which is superior!

Scotch ram	Mourne ram
1. Hollow weak back.	1. Strong straight back.
2. Small hindquarters and pelvis, bad breeder.	2. Large hind-quarters.
3. Short hairy legs no good to climb rocks or for snow.	3. Long, clean slender legs can climb anywhere.
4. Spread toes liable to foot rot, walks on his heels out at the knees.	4. Close toes not liable to foot rot, walks on his toes.
5. Thick lips, whins will give him (?)	5. Straight knees.
6. Spread teeth can't clip grass.	6. Small mouth
7. Big heavy nose impedes sight.	7. Close teeth – good cropper of Grass.
8. Bad balance of head.	8. Long straight face and jaw.
9. Can't reach under stones for grass.	9. Bright alert eye.
10. Horns too far back set, no protection	10. Well set horns will protect him.
11. Too heavy shoulders no use to butcher.	11. Light shoulders.
12. Wool too harsh – bad price.	12. Long, slightly silky wool.[23]

No register has been established for Mourne sheep, but the recent establishment of a Mourne Sheep Breeders Association may lead to more systematic management of surviving flocks.

MANAGEMENT OF SHEEP

In the seventeenth century, some huge flocks of up to 20,000 sheep were recorded in Ireland, and in the 1770s, Arthur Young described large areas as set aside for sheep walks. He recorded some big flocks of sheep. McCarthy of Springhouse, County Tipperary, was said to have flocks of 8,000,[24] while at Brownshill, Mr Brown kept 800 sheep, which ate 25 tons of hay in winter.[25] By 1807, the sheep fair at Ballinasloe was huge. According to George Cully, 'The collector of the tolls told me that there were 95,000 sheep shewn at that time and that often there had been more.'[26] However, overall numbers of sheep in Ireland remained low until the mid-nineteenth century, when a massive increase began, especially in upland areas. Depopulation and a swing away from arable farming provided an incentive for many landlords to introduce large-scale, systematic sheep production in marginal areas. There is little evidence of the brutal clearances that took place in the Highlands of Scotland

– the Famine and emigration meant that many marginal areas in Ireland were cleared already, but there were conflicts between tenants and some landlords who followed the Scottish example and took over huge areas of their estates for sheep. As mentioned earlier, in the 1850s, for example, Lord George Hill of Gweedore reserved 12,000 acres of common land on his estate for his own use. He brought over Scottish graziers and large numbers of sheep. In 1858, ten priests wrote to Parliament protesting about this. 'This fine old Gaelic race is about being crushed to make room for Scottish and English sheep.'[27] Even more controversially, a neighbouring Donegal landlord, George Adair of Glenveagh, evicted 244 people, after conflicts arising over the introduction of Scottish sheep and shepherds to his estate.[28]

By 1900 there were almost 4.4 million sheep in Ireland, and by the 1990s this had risen to 9.9 million.[29] In 1900, the main centres of the sheep trade in Ireland were the September and October fairs at Ballinasloe, the September fair at Banagher, the October fair at Tuam, and the autumn sales in Dublin. Tuam and Ballinasloe are in County Galway, and Banagher is in Offaly, on the border with County Galway.

> The western breeders … who own the lighter lands, breed the sheep and raise them to two and three-year old, and then sell them to the graziers of eastern, midland and southern counties. Either as ewes for breeding purposes, or as wethers to be sold off … the counties of Galway, Mayo and Roscommon, furnish much the largest proportion of these store sheep.[30]

The counties with most sheep in Ireland reflected this trade to a large extent.

Number of sheep in 1900, by county

Galway	653,456	Meath	234,676
Mayo	361,978	Wicklow	228,820
Cork	320,361	Wexford	208,423
Tipperary	251,202	Roscommon	192,459[31]

In the 1770s, Arthur Young had found some evidence for foddering sheep, on hay, turnips, and cabbage. At Ballybar, County Carlow, for example, a Mr Butler bought wethers in October, and sold them in spring and summer. In the winter the sheep were fed on hay,[32] while at Belleisle, County Fermanagh,

Lord Ross kept a small field of turnips and cabbages for feeding sheep in winter, and found that cabbages were the best fodder, and lasted longest.[33] More developed feeding systems were sometimes used to fatten wethers. At Pakenham, County Kildare, for example, Young found that a Mr Marly fed wethers on bran and oats, a quart of each a day.[34] However, as late as 1877, the fattening of sheep in spring by feeding them on roots was found only on some Irish farms. Hay or roots were recommended for store sheep and breeding ewes in severe winters.[35]

At many places, including Kilfaine, County Kilkenny, Courtown, County Wexford, and Mount Kennedy, County Wicklow, Arthur Young found that folding was not practised, and at Mount Kennedy there was 'no such thing as a hurdle known'.[36] Dean Coote, near Mountmellick, County Laois, told Young that he had folded his sheep for two years, 'but could not bring his people to continue it without too much trouble'.[37] Young pointed out that the use of hurdles allowed farmers to control winter grazing, and meant that farmers would have some fat sheep to sell in the spring. He recommended that Irish farmers should grow sally rods for making hurdles and suggested that Irish labourers should be sent to England for a couple of months, to learn hurdle making.[38] Young recorded one unusual method of restraining sheep near Castleward, County Down. 'Here I saw sheep grazing in a ditch, confined by a line fastened by two pins, and drove into the ground, and passing through rings which hung from a strap round their necks, so that they could move only from one end to the other.'[39]

Extensive grazing on unimproved land was practised by some big landholders. The 'Great Improver' Lord Chief Baron Forster, for example improved land that had previously been 'a waste sheep walk'.[40] The control of sheep in better land was a problem. In County Derry in 1802, for example, it was claimed that clover and hedge quicks were prevented by growing by 'promiscuous flocks of sheep which are emphatically called *pirates*'.[41] Letting sheep range over a farm could be part of good management, however, even in the twentieth century. John Drennan, who farmed 640 acres of arable land beside Lough Foyle until the 1950s, found that letting young sheep forage over the land could be very beneficial.

> Latterly … I started … [to] buy cross-bred lambs from the hills round here … and in the winter time we just used to let those lambs run all over the farm so that they could clean up the grass and so on, with the

result that in the spring of the year, after you put a good touch [of fertilizer] on the ground you could get good early grass.[42]

Arthur Young observed some well-developed systems of sheep management. At Mount Kennedy, County Wicklow, for example, despite having noted the lack of folding, Young found that there was a significant production of suckling lambs for the Dublin market. (The wool of the sheep was all used locally.) One farmer, 'most able in that business' described his system. He bred his lambs from stock partly bought in each year. He sold about 85 lambs a year. The ewes were made ready for the rams 'by being given ale and let into the fields.' The lambs were born around Michaelmas (29 September). They were left in the field for a week, and then housed. The ewes were brought in between twice and four times a day to suckle them. Later 'they have cow's milk given them by women from their mouths, squirted down the lambs' throats'. The amount of milk given in this way rose from about a quarter of a pint to a half pint.[43] Production of lambs for Dublin markets also developed at other places around the city. At Killadoon, County Kildare, for example, a local landlord fattened lambs for the market at Smithfield, where he believed long legged lambs sold best.[44]

A modified version of this system still operated in the Wicklow Mountains in the 1840s. The lambs were born in December and January. They stayed with the ewes for a fortnight, and were then put into pens in the 'feeding-house', and the ewes were driven in twice a day to suckle them. After a time they were also fed with cow's milk. David Low praised this system, which meant 'that the inhabitants of Dublin are supplied with as fine early lamb as any part of the United Kingdom'.[45]

By the later nineteenth century, farmers in hilly areas often looked to Scotland for expertise in sheep management, and amongst other things, Scottish shepherds introduced new types of sheep fold, and probably sheep dogs. It has been claimed that the use of dogs for herding sheep has been known in Ireland since ancient times. An Irish collie type was recognized in the early nineteenth century,[46] but in the 1840s sheep dogs used in Ireland were described as either English or Scottish, and it was the 'colley' introduced from the latter country that became the most common sheep dog used in Ireland.[47]

Sheep also continued to be kept on a small scale, however. At Keadew, near Burtonport in County Donegal in the 1920s, for example, the Ward family kept six ewes on a shoreline meadow that was regularly flooded at high tide, so ensuring excellent pasture. Each year the ewes would produce up to twelve lambs. These were bought by a local butcher, who claimed that the meat from the lambs was twice as good as anywhere else.[48] In County Galway, Claddagh type sheep grazed on seaweed, while grazing sheep on sea meadows was also seen as beneficial in Gaoth Dobhair, further north in County Donegal, where farmers with rights to grazing pastures on the *machaire*, a sandy area behind the beach, for about three weeks each year. However, if sheep were left longer on the sandy ground, as happened when grazing rights became confused in the 1980s, local butchers found that the stomachs of slaughtered sheep were filled with sand.[49]

Sheep farming on a larger, but still modest scale also continued in areas such as the Mountains of Mourne, County Down. In the 1940s, the mountain pastures were owned in common by farmers on what had been the Hilltown Estate. The land was unfenced, and each farmer paid a small fee for grazing rights. The rights were controlled by a committee of farmers. Some farmers had as few as five or six ewes, but most had about twenty. The lack of fencing was not a problem on this mountain grazing. 'Each farmer knows on what part of the extensive hillside to search for his own black-faced sheep – "Every ewe has her own pasture." … Sheep are marked with pigment or tar (a process known as "keeling") until about one year old – then branded on the horns.'[50] Mourne lambs were born rather late in the year, in April, and sold in fairs in the nearby towns of Rathfriland and Hilltown in September. A dance known as the Tup Dance was organized at the autumn fair at Hilltown, and until at least the 1980s, it was customary to lead a Mourne 'tup' or ram on to the stage at the start of the evening.

SHEEP SHEARING

By the 1830s, it was recommended that sheep in good condition could be shorn in May or early June.[51] Systematic washing of sheep before shearing was well established by this time. John Sproule recommended that washing should take place about eight days before shearing.

> The washing pool may be in any part of a stream or rivulet, from two to three feet in depth, and the bottom should be pebbly rather than muddy … [The] sheep are brought to one of the banks of the rivulet … Three or four men will be necessary to stand in the water … The sheep is handed to the first man, which he seizes and pulls into the water, and

immediately turns over on its back ... He is thus enabled to turn the sheep over from side to side, while at the same time he pulls it gently backwards and forwards ... In this process the wool waves up and down the length of the body, and swirls round ... first in one direction and then in another. He then hands the sheep to the next washer, who repeats the same operation, and he to the next, and so on to the last person ... By this operation ... the skin and wool of the sheep are cleaned as if they had been washed in soap and water; and, in fact, they have been so washed, for the oil or yolk of wool presents similar chemical properties to soap.[52]

Dipping sheep in disinfectant, rather than simply washing them, was becoming widespread by the end of the nineteenth century (fig. 116). In the early twentieth century, the Congested Districts Board claimed credit for introducing portable sheep dips to western areas, such as Clifden and Doon in County Galway, and Glenties in County Donegal. In the year 1903–04, 130,000 sheep were brought for dipping at troughs supplied by the Board.[53]

Sproule recommended that sheep should be sheared on a winnowing cloth to keep the fleece clean. The shearer should set the sheep on its rump, and keep it in that position with its back resting against the shearer's legs. The wool was removed from the head and neck first, and then from the belly towards the back. The sheep was then laid on its side, kept in position by the shearer's leg. The shears should be held as closely as possible to the sheep's body. Short, narrow strokes were recommended, to make a clean clip. The clip marks should appear as concentric circles. After shearing, all dirty portions around the tail and belly should be removed. These could be washed separately, and used as stuffing in saddles and horse collars. The fleece itself was folded inwards from the sides, and then rolled up. The wool from the neck was then stretched out and twisted into rope, which was then wound round the fleece.[54]

At Annsgrove, County Cork, Arthur Young claimed that all the cottars had sheep, which they milked for their families,[55] but this seems to have been rare. The vast majority of sheep were kept for meat and wool. At Belleisle, County Fermanagh, for example, Young found that lambs were sold at three to four months old, for around 7 to 8s. each. The wool of a ewe fetched 4s. 4d. Some farmers bought wethers in the month of June, for fattening, and sold them the following spring for between £1 1s. to £1 6s.[56] Around Florencecourt in County Fermanagh, however, Young found that 'the little farmers that have land fit for

116 Sheep dipping, County Antrim, *c.*1910 (UFTM, WAG 1945).

sheep, keep a few for cloathing their families … They clip from a ewe, about 3lb. on average.' He recorded fleeces as light as 2½ lbs, but some fleeces were much heavier. At Courtown, County Wexford, which was 'not a sheep country,' a farmer who kept eight ewes on two acres got a stone of wool from each of them.[57]

Near Mountmellick, County Laois, the wool trade was more commercially organized. The wool was bought by 'combers' who kept spinners in the country. The Dublin Society sold fabrics made from Spanish, South Down and Wicklow wool, which was claimed to be equal in texture and durability to any English cloth. In the early nineteenth century, weaving factories were established at locations such as Ennisnag, County Kilkenny. Water power allowed large-scale manufacture of woolens from the middle of the nineteenth century onwards. For example, in the 1890s, the factory at Athlone employed around 400 people.[58]

LAMB AND MUTTON

The consumption of mutton and lamb grew in the eighteenth century. In general, it was not salted, and it was never so widely eaten as pork or beef.[59] There was only a small export trade in mutton and its home consumption made economic sense. In 1877, Thomas Baldwin claimed that the longer a sheep takes to come to maturity, the better the quality of its mutton, so slow growing unimproved sheep had the best meat.[60]

John O'Connor, who was born near Screen, County Wexford in 1911, remembered sheep, mostly Lincoln and Border Leicester, being divided between neighbours.

> Sheep got plentiful, [so there was] no market for them [so they would kill one] – hung up the same way as a pig. [and] divided up that night between four or five houses. Then another got killed a couple of days later. [They] used up two sheep a week that way. When the price went up, you got no mutton. [This] happened every ten years or so.

John O'Connor also remembered mutton being pickled in Wexford, in the same way as pork.[61] A nineteenth-century description of dry curing mutton also shows similarities to dry curing pork.

> Good ham may be made from any part of a carcass of mutton … for this purpose it should be rubbed all over with good salt, and a little saltpeter, for ten minutes, and then laid in a dish and covered with a cloth for eight days. After that, it should be slightly rubbed again, for about five minutes, and then hung in a dry place, say the roof of a kitchen, until used. Wether mutton is used for hams, because it is fat, and it may be cured any time from November to May; but ram mutton makes the largest and highest flavoured ham, provided it be cured in spring, because it is out of season in autumn.[62]

By the early twentieth century, most mutton was processed in factories. In the 1920s, Matt Lyons of County Leitrim began shipping beef quarters and lamb carcasses to England.[63] By the end of the century, three factories, Kepak, Dawn and Slaney Meats dominated the production of sheep meat, with 75% of production exported to France.[64]

CHAPTER SIXTEEN

Pigs

PIGS ARE VALUED as a relatively cheap source of meat, easily fed on household refuse and natural food supplies, such as autumn leaf mast. In Ireland from the late seventeenth century, when the potato became the basic fodder for pigs, they were a way to 'store' excess produce of the crop, and pigs came to be known as 'the gentleman who pays the rent', and as a 'savings bank'.[1]

In 1780, Arthur Young was amused by the number of pigs in County Cork

> Hogs are kept in such numbers that the little towns and villages swarm with them; pigs and children bask about, and often remember one another so much, that it is necessary to look twice before the human face divine is confessed. I believe there are more pigs in Mitchelstown than human beings and yet propagation is the only trade that flourished here for ages.[2]

Pigs were particularly associated with small farms.

> The number of pigs in Ireland, in 1841, was 1,412,813
> Of these there were owned by persons holding under one acre … 355,977
> By those holding from one to five acres … 254,437
> By those holding from five to fifteen acres … 342,436
>
> Making a total of 952,850 pigs owned by those holding under fifteen acres each, and only 459,963 by those holding over fifteen acres.[3]

The close relationship between pigs and potatoes became very obvious during the Great Famine of the 1840s, when the failure of successive potato crops also led to a drastic fall in pig numbers, from 1,412,813 in 1841, to 565,629 in 1848. Numbers recovered rapidly after the Famine, rising to 1,084,857 by 1851, but because of the association of pigs with small farmers and cottiers, the loss of pigs was particularly devastating for the poor.[4]

117 Irish Greyhound pig, *Irish Farmer's Gazette* (1847), p. 57.

IRISH GREYHOUND PIG

The best-known type of pig in nineteenth century Ireland was the Irish Greyhound pig (fig. 117). These were described as, 'a very rough race of animals – colour white, standing very high on the legs, with narrow carcasses, thick coarse bone, rough bristles and hanging ears.'[5] Sporadic attempts to improve the native animals were made in Leinster and Ulster throughout the nineteenth century. Some experts were confident that the Greyhound pig could be improved. In 1813, Horace Townsend argued that many of the defects attributed to native Irish pigs were not inherent, but rose from their care and management, and particularly from malnutrition when the animals were young. 'Every poor housekeeper as well as farmer, keeps at least one pig, and neither the former nor the latter has any idea of giving the creature a bellyful until a few months before it is thought necessary to fatten him for market.'[6] The Greyhound pig's pork was praised.

> The native Irish hog … fattens to an enormous size, and makes the best mess pork in the world, being the only salted meat that remains good in

118 Improved Irish pig exhibited by a cottier, Peter Flood, at the RDS Show in 1847, *Irish Farmer's Gazette* (1847), p. 57.

all climates. The author of *Cook's Voyages* relates, that the Irish mess pork was found perfectly sweet and good, at the end of a five years' voyage.[7]

The Greyhound pig was relatively easy to improve when properly housed and fed, but also when it was crossed with introduced breeds. A striking example of this improvement was recorded by H.D. Richardson at the Royal Dublin Society Show in 1847 (fig. 118). The pig was the property of a cottier called Peter Flood and was thought to have been a cross between a Greyhound pig and a Hampshire. It weighed over forty-one stones and was said to have 'put up flesh' with a very small portion of feeding.'[8]

ENGLISH BREEDS

Improved English breeds of pig were becoming common in Ireland by the 1850s. In 1852, for example, the Revd John Warburton of Kill, County Kildare, won a number of prizes at an agricultural show in Galway, for pure-bred Berkshire sows and a boar.

'Large Yorkshire' pigs also began to be imported to Ireland in the 1850s. An important step in pig improvement was taken at the Albert Institute,

Glasnevin, County Dublin in the 1860s, when Prince Albert sent the institute several 'Improved Yorkshire' pigs from his herd at Windsor. The foundation of the Glasnevin herd was laid by crossing these Windsor pigs with the best animals from Irish herds. By the 1880s, the pigs at Glasnevin possessed all the characteristics of the best strains of the Large White Yorkshire pigs. The object was to produce animals that would grow quickly, with a minimum amount of offal. By 1902, it was claimed that all the stock pigs in the herd were 'remarkable for their even-fleshed bodies, good hams, straight legs, thin skins, and large quantity of silky hair. The herd is kept in a normal breeding condition, and none of them are made up for show.'[9]

In the 1870s, some of the bacon curers in Munster made efforts to improve the pigs in the districts from which they drew their supplies, and in the 1880s, an organized effort was made to persuade farmers to breed pigs that were suitable for the production of high class bacon. Munster had taken the lead in the bacon curing business, probably because it had the best dairying districts in Ireland. Dairying and pig keeping were closely associated, as buttermilk was an excellent food for pigs.

However, a large number of the pigs cured in Munster were bought in Connacht, where the pigs were said to remain poor in quality, bad in shape, and black in colour. Boars of the Large White Yorkshire breed were imported and sent to remedy this, but it was claimed that for a very long time farmers were reluctant to use the boars.

> They still clung to the long-legged, flat-hammed animal, whose unthriftiness was in sad contrast to his appetite, with the result that for years the prices quoted by the bacon merchants for Connaught pigs were always a couple of shillings per cwt. under the prices quoted for those in Munster. Perseverance eventually conquered, and to-day as fine pigs can be found in Connaught as in any other province.[10]

In 1902, the Large White Yorkshire was still seen as the best breed for improving native Irish pigs. The 'nearest approach to perfection' was thought to be 'something between the Large and Middle York breeds'. The Berkshire, although an excellent breed and very popular with English curers, was said not to have 'nicked' with the common Irish pig in Leinster or Munster, and the Tamworth and Suffolk breeds were not seen as serious competitors. The cross between boars of the Large White Yorkshire breed and native sows was

119 Large White Ulster pig (UFTM, L832/9).

successful both from the farmer's and the curer's point of view. The pigs resulting from the cross had a good constitution, good digestion, grew quickly, and bred prolifically. They suited the curer because the more valuable cuts predominated in the body, and the offal was light.[11]

However, beyond what had been achieved by the Bacon Curers' Association, in 1902 it was claimed that little other effort had been made to encourage the breeding of good quality pigs, compared with what had been done for horses, cattle, and sheep. Labourers and cottiers, whose pigs represented their largest investment in live stock, and also the foundation of the country's bacon curing industry, were hardly reached by the annual Royal Dublin Society shows, or the limited number of shows held elsewhere in the Ireland. Even where pigs did appear at shows, it was claimed that prizes were awarded to animals that were fatter than their competitors.

The herd book started by the Royal Dublin Society around 1900, and the scheme of Service premiums established by the Department of Agriculture were seen as a way forward.[12] In 1902, the South of Ireland Bacon Curers' Pig Improvement Association had three breeding establishments, one at Limerick, one at Cork, and another in Waterford. An inspector was attached to each of these, whose duty was to keep in constant touch with local boar-keepers, and supply them with boars bred at one of the breeding establishments, or purchased from reliable breeders. It was hoped that these boars would rectify faults in the pigs bred in the districts where they were stationed, and would also prevent in-breeding. The Congested Districts Board also distributed boars in the west of Ireland. By 1902, Donegal had received 37, Cork 15, Kerry 15, Mayo 83, Galway 40, Sligo 7, Leitrim 15, and Roscommon 15. The Department of Agriculture issued its first scheme for the improvement of the breeding of swine in May, 1901. The Department believed that offering premiums would have the effect of inducing more farmers to go in for breeding pure-bred animals.[13]

LARGE WHITE ULSTER PIGS

Selective breeding with imported pigs, and (despite concerns raised in southern Ireland in 1902) especially the Berkshire, led to one of the most successful of all Irish farm livestock breeds, the Large White Ulster (fig. 119). As elsewhere, the breed may have been partly developed using Greyhound pigs. In 1923, R. Wallace suggested that the harsher climate in the north had led to an indoor system of breeding and feeding which had the effect of promoting early maturity, and made the hairy coat of the Greyhound softer and finer.[14]

In 1812, John Dubourdieu recounted that in County Antrim ten years previously, a Revd Dr Percy and a Revd Wm. Moore had introduced Berkshire pigs. Local farmers were said to have 'had recourse' to the offspring, for which they were said to have paid very high prices. Dubourdieu claimed that crosses between the Berkshire and native pigs had become 'very general throughout the northern counties', to the extent that they 'nearly superseded the native long-legged flat sided animal'. The cross with the old breed was 'thought to be of general utility'. It grew faster than the Greyhound pig and the females were more prolific and better provided with milk. These qualities of early maturity, fecundity and the ability to provide an abundant supply of milk were qualities of the modern Large White Ulster breed.[15]

Dubourdieu also reported the upgrading of the Greyhound pig by local farmers in County Down. He said that the native pig was improving into 'a more compact and robust creature, that not only takes flesh faster but is also able to carry its flesh'. The imported breeds here were the Berkshire and the Dutch.[16]

During the nineteenth century there was no centrally organized programme of improvement in the north of Ireland. Despite this, by 1902, it was recognized that farmers by taking the matter in their own hands had 'succeeded in maintaining a fair standard of quality in their pigs'.[17] The uniformity of the type of pig produced by local farmers, its widespread establishment in the north and its ability to breed true to type, gave the pig the requisite characteristics of a distinct breed.

To ensure that the standard of pig produced by Ulster farmers was maintained and improved, the Royal Ulster Agricultural Society, after consultation with breeders and bacon-curers, recognized the pig as a pure breed. In 1907 a Large White Ulster Herd Book was established, with a standard description and scale of points. Inspection for registration was to be made at various centres throughout Ulster at a given period each year.

The ears were a distinctive feature of the Large White Ulster, hanging down over the eyes and snout. Supporters of the breed argued that the long drooping ears were a vital point on the pig as it contributed towards a quiet nature and adapted it for field grazing. In the Bourdelle classification of pig types, 'well developed pendulous ears covering the eyes' were described as a feature of the Celtic type of pig. Both the Irish Greyhound pig and the Large White Ulster had this distinctive feature.[18]

The Large White Ulster pig was the principal breed reared in Ulster during the first half of the twentieth century. It was claimed that 'for early maturity and economy of production, high class flavour and quality of bacon' the Large White Ulster was unsurpassed.[19] The large White Ulster was also fat, which was regarded as a sign of a good bacon producing pig, when there was little demand for a high ratio of lean meat to fat.

A number of factors have been suggested as contributing towards the demand for fat pigs. Before refrigeration, climatic conditions and slowness of transport to distant overseas markets, necessitated a hard cured bacon. Hard cured bacon contained a high percentage of salt, and because heavy salting made lean meat relatively unpalatable, and this led to the demand for a very fat type of bacon.[20] Another factor suggested as contributing towards the demand for fattish bacon was the greater amount of manual labour performed in the past. The amount of fat a person can eat varies more or less in proportion to the amount of muscular energy employed.[21] One of the major export areas for Ulster bacon was the mining districts in the North of England, where it was claimed, fattish bacon was particularly appreciated.[22] However, although the unique fattening qualities of the Large White Ulster were praised, the side effects produced by its facility to fatten sometimes impaired the physical capabilities of the animal. One County Antrim farmer said that, 'there was so much fat about the head they nearly all went blind, they couldn't see where they were going'.[23]

Despite their tendency to fatness, the quality of the bacon produced by the Large White Ulster was widely commended, particularly when compared with modern bacon. A favourite characteristic referred to was the rich aroma from the bacon when it was fried.[24]

Other characteristics which made the Large White Ulster popular with farmers were its breeding qualities. It was estimated that the average litter produced by a Large White Ulster sow was a dozen, but litters of up to twenty-two piglets were recorded, and as well as being prolific the sows were good sucklers.[25]

Large White Ulster pigs were still common during the 1930s, when the Irish Department of Agriculture recognized the importance of the breed in the border counties of Cavan, Monaghan and Donegal.[26] However, by the mid-twentieth century, the Large White Ulster pig had declined in numbers. During the nineteenth century and first half of the twentieth century, Ulster farmers geared their pig production towards the rolled bacon and ham trade, developing the Large White Ulster to suit that trade. The main export markets for this trade were the north of England and the industrial areas of Scotland. However, by 1934 the Ministry of Agriculture for Northern Ireland found that despite a growing pig population, the export market for the rolled bacon and ham trade had become 'relatively restricted'. On the other hand, there was an increased demand for lean Wiltshire-cured bacon among the growing population of London and the south or England.[27] The Wiltshire cure required a relatively immature pig of about two hundred pounds live weight with a large amount of lean to fat. Consequently, any expansion into the Wiltshire market would necessitate a change in the breed of pig kept by Ulster farmers.

The Northern Ireland Ministry of Agriculture set up a Pig Marketing Scheme in 1933. As a direct result of this factories for the Wiltshire cure were established.[28] In order to preserve a standard of quality which would accommodate different markets, Ulster farmers had to change from the Large White Ulster to a breed of pig suited to both the traditional bacon and ham trade and the Wiltshire trade. Between 1934 and 1939 there was a change in breed from the Large White Ulster to the Large White York, which as we have seen, was already established in most parts of the south of Ireland. The Large White York was suitable for the traditional roll bacon and ham curing as well as the Wiltshire cure.[29]

The change happened very quickly. Between 1933 and 1934, the number of registered Large White Ulster boars declined from 142, to 86, while the number of Large White York boars in Northern Ireland rose from 110 to 238.[30] By the 1960s, the Large White Ulster pig had become extinct.

CARE AND MANAGEMENT OF PIGS

As we have seen, the dependence of both humans and animals on potatoes amused Arthur Young in 1780. By the early twentieth century, however, the notion that pigs and humans could share the same food was a cause of understandable resentment. The Donegal writer, Patrick MacGill described his food on a small farm in County Tyrone, where he was hired at the age of fourteen.

> In the morning I was called at five o'clock and sent out to wash potatoes in a stream near the house. Afterwards they were boiled in a pot over the kitchen fire, and when cooked they were eaten by the pigs and me. I must say that I was allowed to pick the best potatoes for myself, and I got a bowl of buttermilk to wash them down. The pigs got buttermilk also. That was my breakfast during the six months. For dinner I got potatoes and buttermilk, for supper buttermilk and potatoes. I never got tea in the afternoon. The Bennets [his employers] took tea themselves, but I suppose they thought that such a luxury was unnecessary for me.[31]

Potatoes are excellent food for pigs because they are high in carbohydrates, which produce hard body fats.[32] However, oats were also commonly used as a supplement to potatoes in the later stages of fattening pigs, especially in the north of Ireland. In County Antrim, for example, potatoes and oatmeal were both used for fattening. Plain oats were also recommended, as 'even though they are not so rapid in their operation, [they] lay the foundation for pork or bacon of the best qualities.'[33]

In 1847, H.D. Richardson suggested that farmers should manage pigs intended for pork and those kept for bacon in different ways.

> Pigs designed for pork should not be fattened to the same extent as those designed for bacon ... Porkers should be suffered to run at large. Grazing, or the run of a wood in which roots or nuts may be met with,

is calculated in an eminent degree to improve the quality of their flesh. Of course it will be necessary to give the pigs regular meals, independent of what they can thus cater for themselves … Bacon pigs fatten best by themselves; they need no liberty; and it is only necessary to keep the sty dry and clean, and to feed abundantly, in order to prepare them for the knife … there cannot be a more gratifying sound to the ears of a zealous pig-feeder than that peculiar, self-satisfied, contented grunt, with which the huge hog, basking, perhaps, beneath a summer's sun, announces to his admiring owner that all his wants and wishes have been satisfied.[34]

Richardson disapproved of feeding pigs on offal and carcasses of other dead animals, which he alleged was common in urban areas in the 1840s.

Pork butchers, resident in large towns, are very apt to feed chiefly on offal, including that arising from the pigs daily slain and dressed for the market. To make swine feed upon the entrails … of their brethren, strikes me as revolting … There is yet another [unpleasant] description of feeding … in knackers' yards. The animals are kept by these persons in large numbers, and are fed wholly upon the refuse of dead horses … I have frequently been disgusted by the sight, in one of these yards, of three or four fierce, wolfish looking hogs, their muzzles plunged to the eyes in the abdomen of a slaughtered horse, and their savage jaws dripping with gore … Pigs are not now so generally kept in Dublin as formerly … yet I would venture to affirm, that a visitor to 'Red Cow Lane' would still find … half a dozen or more foul-feeding and strong-smelling swine, banqueting on the corrupting carcass of some wretched old horse.[35]

Pigs flourish in the same temperature conditions as people, and the lack of purpose-built outbuildings on many farms led to the practice of keeping pigs in the farm kitchen, especially sows when they were farrowing. This persisted in some areas until well into the twentieth century. In the Glens of Antrim, for example, one farmer interviewed in the 1980s, recalled that in one house where he socialized, the sow would be put behind a settle bed in the kitchen, and the piglets, when not suckling their mother, were put in a tea chest nearby. The sow would sometimes grunt from behind the bed, making the piglets squeal for food. Men holding a bottle of stout in one hand, would each lift out a piglet, bringing them to their mother, so that 'they got their bottle too!'[36]

At the same time, however, large-scale pig farming was becoming increasingly sophisticated. In the 1940s, the Jordan family's farm at Kilmore, County Armagh, for example, had over 2,000 pigs. This increased to around 25,000 pigs by the 1980s. Up to 75 pigs were kept together in 'sweat boxes' to keep to keep them from getting too fat. During the mid-twentieth century, systems were installed that allowed up to 12,000 pigs to be fed in an hour, and slurry was brought to the fields by up to three-quarters of a mile of pipes.[37]

SLAUGHTERING AND CURING

A statistical review of the Irish Bacon and Provision Trade by the Solicitor-General, in 1860, summarized the development of Irish pork trade with London and the British colonies.

> During the Peninsular War Ireland possessed a great trade in curing beef and pork. Cork, Waterford, Limerick, and Dublin, all afforded their quota of beef to the English navy. Upon the proclamation of peace this trade fell off greatly, and the introduction of steam navigation, in 1825, tended still further to diminish the trade, for thus a ready market was opened in England for the live animal. Again, the repeal of the laws prohibiting the import of foreign cattle and provisions still further affected this trade … Live animals and bacon now form the staple article of the Irish provision trade … Waterford produces nearly two-thirds of the Irish bacon imported into London, and the pigs supplied by the adjacent counties, Waterford, Kilkenny and Wexford, not being sufficient for the wants of the trade, Waterford buyers attend the fairs in Carlow, Tipperary, Cork, and Limerick, extending their journeys at times into the midland counties, into Connaught. If pig-feeding be, as no doubt it is, profitable to farmers, it follows that facility of access to the principal market is of great importance to them.[38]

The slaughter of pigs on farms, for both home and commercial use, was common in many parts of Ireland. In 1847, Richardson described a common method of slaughtering.

> In country districts of Ireland, the pig is usually secured by the hind leg to a post of ring, the head is fastened to another; the animal is thus securely strapped down upon a sloping slab or table, and the head is severed from the body by means of a knife.[39]

120 Slaughtering a pig in County Armagh, *c.*1900
(F.J. Bigger collection, Ulster Museum).

In the later nineteenth century, some slaughtering methods were described as 'primitive' (fig. 120).

> The pig having been stunned by one or more blows of a mallet, as the case might be, its throat was cut and the blood allowed to flow. The carcass was then surrounded by a quantity of straw or reed, which was set on fire in order to burn the hair off the skin, which was then scraped after hot water had been thrown upon it and it had been hung up by the hind legs. Having been disemboweled and left suspended in the hanging house for twenty-four hours, it was weighed and paid for as dead weight.[40]

Some pigs were slaughtered in this way until well into the twentieth century, when a pig was slaughtered for home use. In Ulster it persisted as the main technique of slaughtering all pigs until the 1950s. The pig killer had to stalk the pig, and when it was suitably positioned, hit it a hard blow on the forehead. As the stunned pig fell to the ground another man got astride it and held it up by the forelegs to enable the pig killer to slit its throat with a knife. The position of the helper when the pig killer was delivering the blow with the mallet was potentially dangerous. If the pig made a sudden move and the pig killer miscalculated his blow, the helper, if not alert, could be struck by accident. In some areas, the local pig killers overcame the problem of moving pigs by using a vicious instrument known as a cleek, which was later made illegal. The cleek was a steel hook which was stuck in the jaw of the pig to hold it steady while the pig killer stunned it with a mallet.[41]

Slaughtering of pigs on farms in Ulster persisted because of type of pig kept, and the system of marketing which differed significantly from the rest of the country. A central feature of the system was the marketing of dead pork rather than a live pig trade. Bringing dead pigs to market for sale appears to have been common in the north throughout the nineteenth century. The Revd Thomas Gaffikin, J.P., remembering Belfast in the 1820s, recalled that, 'in Tomb Street the weighing market was so much in requisition, that on a winter's morning all the streets approaching it would be blocked with carts loaded with dead pigs and firkins of butter.'[42]

A major factor leading to the farm slaughter of pigs and the dead pork trade was a characteristic of Large White Ulster pigs; their thin skin, said to be a feature 'much desired by the high-class bacon curer'.[43] A thin skinned pig bruises easily in transit and any decline in the physical state of the pig reduces its marketable value. In 1932 a Northern Ireland Government Report, discussing pig production in the north and the viability of a live pig trade, concluded that 'The typical Ulster pig is not one that will drive easily and the loss in condition occasioned by a fruitless visit to the market and subsequent return to the farm would probably involve very serious deterioration in the marketable condition of the pig.'[44] This made slaughtering on the farm 'almost a necessity'.[45]

Generally, pigs were killed when the farmer thought it would weigh 1¾ cwt dead weight. Slaughtering was usually carried out by a local butcher or pig killer. It usually took place the day before the pig market and required the use of extra labour besides the pig killer. This included arranging for help in catching and carrying pigs, and having an adequate supply of boiling water available for scraping the carcasses.[46]

After the pig was killed and bled, the next stage in processing the carcass was to scald it in a tub of very hot water. This was done to remove the hair from the pig's skin. The temperature of the water was crucial. If the water was too hot, the skin of the pig could be damaged, particularly when the pig was thin-skinned, like the Large White Ulster. During the nineteenth century it was thought if the pig was scalded before it had completely expired, the hair was easier removed. Richardson, complaining of the cruel practice remarked:

> In the country districts of both England and Ireland … old abuses are still permitted to exist; with a few honourable exceptions, the barbarous practice of plunging the pig into scald, while yet living is still systematically and designedly adopted. A very respectable man surprised me the other day, by deliberately telling me that 'A pig will no way scald so well as when the life is in him'.[47]

After the pig had been thoroughly immersed in the scald and the hair loosened from the skin it was lifted out onto a board, the hair shaved off and the heels of the back legs slit. A strong stick was run through the track cut in the heels of the pig, and after the hair had been shaved off it was lifted and hung by the heels in a suitable place, where the pig killer could split and disembowel the carcass.

Because pigs in the north of Ireland were killed and cleaned on the farm, the 'pluck' or offal of the pig in Ulster was utilized as food at the farm house more frequently than in other parts of Ireland. The main parts of offal used were the heart, liver, and the inside fat. The stomach might be used for tripe, or stuffed with potato, oats and scratchings, boiled, sliced and fried in an Ulster version of haggis, sometimes known as 'haddock'. The use of intestines for sausages and puddings was known throughout Ireland. Black pudding, white puddings, and sausages were all made on the farm if a pig was killed for home use, well into the twentieth century.

After the innards were removed from the carcass, it was washed clean and left to cool for market the following day. The carcasses were normally transported to the local pig market by horse and cart. On arrival, the general procedure was to line up in the market yard or street and wait in the queue at the weighing scales.[48]

In the north, however, by the 1950s, the killing and curing of pigs at abattoirs was almost as common as elsewhere in Ireland. The home slaughter and curing of pigs in Ulster had almost disappeared in 1954,[49] and is now illegal both north and south.

FACTORY CURING

Factories for curing pork and bacon were well established in cities and towns during the nineteenth century. Alexander Shaw claimed that according to tradition the birthplace of the bacon curing industry was Baltinglass in County Wicklow. Wicklow had a large number of small curers, who cured long sides for the Dublin market. This particular cut of bacon was still being turned out in 1902. Shaw said that the greatest impetus given to bacon curing was undoubtedly the rapid advance made by the dairying industry in Ireland generally, and especially in the province of Munster. One of the best ways to utilize the waste products of the dairy was in pig-feeding, and consequently the pig became a necessary adjunct to every dairy farm. The largest curing centre in Ireland was Limerick. In 1902, the annual turnover there was almost equal to that of Cork and Waterford together. After these cities, Belfast was the biggest centre, then Derry, Dublin, Tralee, Enniscorthy, Dundalk, Ballymena, and New Ross.

In Limerick curing establishments, as on Ulster farms, the bristles of slaughtered pigs were removed by scalding, previous to curing. However, in 1902, pigs intended for bacon for London had the bristles taken off by singeing. It was claimed that bacon prepared in the former way would not sell in the London market. Belfast bacon and hams were shipped in 'a finished condition,' dried and smoked, while bacon and hams from the south of Ireland, with the exception of some of the meat manufactured in Limerick, were shipped in an undried state, and were dried and smoked on arrival in Britain. The bacon sent to London differed from that cured in the north in another way. The ham was not separated from the flitch, but was shipped in 'bales', each bale consisting of the flesh of two pigs.[50]

Before the advent of steam ships, it was necessary that bacon should be sufficiently salted to survive a longer voyage. However, after a regular weekly line of fast-sailing ships was established between Waterford and London, sailing once a week, the curers could moderate the amount of salt used. Once steamships were introduced, care was taken to prevent over-salting, and the resulting bacon was claimed to bring the highest price in the best markets in the world.

In 1902, Alexander Shaw wrote that 'the hard cured bacon of former days would now be looked on as akin to Lot's wife, and it was by mere chance that the change in taste was brought about'. In 1880, he claimed, 'a struggling

Limerick curer, being on an occasion unusually short of money, in order to turn his bacon into cash was obliged to turn it out in what was then considered a half-cured condition. However, the people who bought this bacon liked it, and asked for more.' The other curers soon followed their neighbour's unintentional lead. 'The manufacture of mild cure has now been brought to such perfection that it can be sent into tropical climates for consumption within a reasonable time.'[51]

The offal produced by slaughter was also an important part of the trade. Sausage and pudding making employed a large number of women, while tinned meats, such as brawn, ham and chicken, etc., were also made, principally for export. Most of the livers were shipped to Germany, to be made into liver sausages.

Until the mid-nineteenth century, curing was generally suspended about the 1st of May, and resumed about the beginning of October. Most of the men employed in curing were only engaged for the season. Several methods of curing bacon in summer were tried but these failed, until a Waterford curing establishment discovered a method of applying ice in the process around 1860. The flitches of pork that were cured using ice were carefully piled in large tanks. Pickle, which had been brought to the required temperature by the use of ice and salt, was poured in. In warm weather, the process was more difficult and more expensive, than during the cooler part of the summer.

Early methods of ice-curing was carried on in a simple way, the ice being left in open crates in the centre of the building where the curing was in progress, in order to keep the air cool. The Harris Patent Ice House improved on this. It consisted of large chambers on iron floors, supported by heavy beams or uprights that could support as much as 1,000 tons of ice at a time in the bacon curing season. The bacon was piled in cellars underneath these chambers and the cold air from the ice overhead descended through the iron floor and kept the temperature low during the summer months.

About 1887 the system was again revolutionized, by the introduction of machinery for the production of cold by a process using ammonia or carbonic acid. The initial cost of these systems was heavy, but it proved profitable, as the work was done much better and at running costs that were half the cost of the old methods. These modern refrigerating plants were well adapted for use in connection with the production of the mild cured bacon.

Ice-cured bacon was described as sound and firm, and, consequently much prized. The development meant that Irish farmers could now market pigs throughout the year, and this could mean increased profits. Pigs require a good deal of warmth while fattening in winter, and this warmth had to be produced by food, so that a much greater quantity of fodder was necessary to bring up a pig to a given weight in winter than in summer. The introduction of a method of curing which enabled fattening in summer was also good for workers, who now could now hope for constant employment. Farmers in the south of Ireland were the first to avail themselves of the opportunities offered by a summer market. The consumer also benefited, in having delicious mild food all year, instead of highly salted bacon.[52]

As in Northern Ireland, intensive pig production developed in the Irish Free State during 1930s. During this decade, for example, Mitchelstown Co-Op engaged a well-known agriculturalist, Sandy McGuckian from Masserene Park, Ballymena, County Antrim to help train local people in modern intensive animal production methods. As a result, several of Ireland's largest industrial pig production units are now based in the Mitchelstown area. In the Republic as a whole, the largest processors of pigmeat are Glanbia, along with Dawn and Dairygold (Galtee).[53] Pig production remains an important part of Irish farming. In the 1990s, more than 2,000,000 pigs were kept on farms, north and south.

CHAPTER SEVENTEEN

Poultry

POULTRY HAD BECOME very common on Irish farms by the time of Arthur Young's tour in the 1780s.

> Poultry in many parts of the kingdom, especially Leinster, are in such quantities as amazed me, not only cocks and hens, but also geese and turkies; this is owing probably to the circumstances; first to the plenty of potatoes with which they are fed, secondly, to the warmth of the cabins, and thirdly to the great quantity of spontaneous white clover (*trifolium repens*) in almost all the fields; upon the seeds of this plant the young poultry rear themselves.[1]

In the 1840s, Martin Doyle said that twenty-two varieties of poultry could be distinguished, but in his work he deals only with the 'barn-door fowl' and the 'mongrel dunghill'. He described the barn-door cock as having a thin comb, an arching tail, long neck feathers, a long neck, and variegated colouring. The best dunghill fowls were said to be medium sized with dark plumage and white legs.[2] By the later nineteenth century, however, there was more interest in distinct breeds of chicken (fig. 121). Thomas Baldwin recommended three breeds especially.

> The *Dorking.* This had dark grey plumage, a square, well-set body, short whitish legs, and a short neck (fig. 121a).
> *Spanish fowl.* These were black, with a white face and a lobe of white flesh behind the ear.
> The *Brahma.* This breed could be either white or dark. It was larger than the other two breeds, but its flesh was said not to be so good. However, it gave eggs all winter, and Baldwin thought it the most useful common breed (fig. 121b).[3]

By the early twentieth century, breeds such as Leghorn (fig. 121c) and Sussex chickens had become popular, while the Rhode Island Red became the favourite breed of many small farmers, as the most successful dual-purpose breed, producing up to 300 eggs a year, but also valued for its meat.[4]

a   **b**   c

121 Common breeds of poultry: (a) Dorking; (b) Brahma; (c) Leghorn, taken from Irish National Schools, *Introduction to practical farming* (Dublin, 1898), pp 226–9).

CARE AND MANAGEMENT

The requirement for specialized poultry houses was not seen as a priority, until well into the nineteenth century. In 1832, for example, John Sproule commented 'When a few poultry only are kept, it may not be necessary to erect a structure for their sole use, as they generally take shelter themselves in some of the sheds, and pick up the greater portion of their food about the barn and stable.' If thought necessary, small scale accommodation could be provided for ducks, geese, hens and turkeys, which could roost on the floor, perches or roof timbers.[5] By 1863, however, instructions were more specific, although it was still emphasised that poultry houses need not be expensive or elaborate. It was recommended that there should be an earth floor, with some loose sand for the poultry to roll in. The roosting perches should be made as a wide ladder, set in a sloping position against the wall. The perches should be less than two inches in diameter, and about one foot apart.[6]

By the 1870s, the specifications for good poultry houses had become much more detailed. Thomas Baldwin urged that poultry houses should be at least twelve feet high, 'as air in a low house soon becomes tainted'. He recommended thatch roofing for heat, and suggested that the interior should be lined with straw, hung vertically and pinned to the walls. The straw should be sprinkled with sulphur to prevent it becoming infested with fleas. Cleanliness was seen as essential. The floor of a poultry house should be cleaned two or three times a day, and the manure removed two or three times a week.[7]

EGGS

Martin Doyle claimed that hens, left to themselves, would produce two broods of chicks a year. They would begin to lay in February, and stop at the beginning of November.[8] Hens lay earliest in warm conditions. Doyle pointed out that 'The hen which roosts in a warm nook of the poor man's cabin, lays earlier in the year than the gentleman's hen who lodges in a cold, damp out-house.' He suggested that a stove used for heating a greenhouse or harness room, could heat a hen house at the same time.[9] Hens between two to five years of age were the best layers. Hens which had become broody, and were used for hatching, could be given between nine to fifteen eggs to sit on. Hens which did not become broody could be encouraged to do so, by 'applying stimulating applications – nettles for instance – to the belly'. (Later, Doyle refers to this as barbarous!) The eggs should hatch after three weeks.[10]

There was ongoing debate as to the number of hens that should be served by one cock. Doyle claimed that 'Every old fowl-woman knows that those eggs only are prolific which are produced by constant intercourse with the male, though, for the purposes of the table, they are better without his intervention, as they are more easily preserved in a state of freshness'. The recommended number of hens to a cock varied between six and twenty. Doyle seemed to accept what he described as 'the French house-wife's view' that for rearing ordinary fowl, a smaller number than twenty hens for each male involved the useless expense of keeping more cocks than necessary.[11]

The ideal temperaments Doyle ascribed to hens and cocks, seem very close to those identified as ideal for Victorian ladies and gentlemen! He quoted an earlier English text approvingly, which described a good cock. 'He ought, withal to be brisk, spirited, ardent, and ready in caressing the hens, quick in defending them, attentive in soliciting them to eat, in keeping them together, and in assembling them at night … In making a choice between two cocks … try to make them fight together, and select the conqueror, for … hens, like other females, always prefer the male who shows most courage and wit.' By contrast, Doyle recommended that hens should be of middling size, and 'neither disposed to crow, nor be passionate.'[12]

Martin Doyle believed that fowl 'in their natural state, picking up what they can get at the barn door', were the best flavoured. However, he also described the fattening of fowl in coups, with milk and meal, and even force-feeding through a tube, the mixture made up of milk, meal and hog's lard or kitchen fat. He gave no indication if this practice was widespread in Ireland.[13]

122 Feeding geese, *c*.1910 (UFTM, WAG 1979).

TURKEYS

According to Martin Doyle, turkeys are the most delicate of fowl when young, but they can become very hardy when full-grown. Turkey hens had to hide their eggs from turkey cocks, as the latter would break them. (Doyle claimed that the average turkey cock was 'a bully and a coward.') However, brooding turkey hens were relatively unaggressive, and Doyle believed that any number of them could be kept together, provided the shed was warm, quiet and dark. Chicks hatched after about three-and-a-half weeks. Eggs were the favourite food for young birds, then nettles and parsley rolled into balls, and meal boiled to the consistency of stirabout. After six weeks, they could eat boiled potatoes mixed with their meal. For everyday feeding, adult turkeys could almost feed themselves, if given the run of the yard and fields. Their pickings could be supplemented by some oats or other grain.[14]

GEESE AND DUCKS

Geese required access to a pond. Martin Doyle argued that, given the tendency of geese to fight with other fowl, and that cattle disliked goose dung on pasture, these fowl were best reared near lakes and rivers. On common land, with pools of water, Doyle claimed that the peasantry derived 'great profit' from rearing geese (fig. 122).[15]

In the eighteenth century, geese were kept for their feathers. At Westport, County Mayo, for example, Arthur Young found that living geese were plucked every year. At Ballynogh, County Leitrim, he was told that every goose yielded feathers worth three farthings or a halfpenny each year. He commented that 'They make a dreadful ragged figure.'[16] Martin Doyle also recorded the practice in the 1840s.

> Geese are regularly plucked twice a year, (early in August and at Michaelmas) for the feathers, by wretches who go about for the purpose of purchasing them. As may be supposed , they pull out the feathers in the most rough and rapid manner to save time, without paying the slightest attention to the tortured birds … Humane housewives do not allow these operators to pluck the quills, (which naturally fall once a year) … nor any feathers except those on the belly and sides.[17]

Martin Doyle identified two main types of duck in the 1840s, a dark coloured breed, which he claimed originated from the Rhone, and the Muscovy, which was similar in colour to the common duck, and was a good breeder and easily fattened.[18] By the 1860s, the large white Aylesbury breed was recommended, and the Rouen, which looked like a large version of a wild duck, but was an excellent layer. As with other fowl, adult ducks were largely left to fend for themselves, as they were useful in picking up slugs and other insects. However, if they were being fattened, it was recommended that treacle and chopped mutton fat should be mixed with their meal.[19]

EGG AND POULTRY TRADE

By the 1870s, eggs and poultry were exported from a number of Irish ports, including Wexford, Derry, and Dundalk.[20] Poultry numbers had increased greatly during the second half of the nineteenth century. By 1900, there were more than 18 million chickens in Ireland, more than three times the number recorded in 1850. Even tiny farms in the west had flocks of around 100 fowl. Bodies such as the co-operative movement saw the potential of the trade arising from this. In 1899, the movement estimated that forty hens gave as

123 An egg merchant, County Cork, 1890s, from *Pictures of Irish life* (Cork, 1902).

much profit as one cow. (This was later revised to twenty hens.)[21] However, early leaders of the co-operative movement identified major problems with the system of trade (fig. 123). Catherine Webb, writing about the movement in England in 1904, summarized the situation very clearly.

> The problem amongst farmers' wives all over Ireland had been to hold on to their eggs until they had a sufficient quantity to make it worth-while taking them to market, particularly when prices were going up. The injury done to the trade by this abominable system of 'holding up' eggs was enormous. The Irish egg – under proper conditions the best in the world was sold at the lowest price and was difficult to sell at all.[22]

In 1897, egg merchants in Liverpool and Glasgow issued a circular threatening to refuse to buy Irish eggs unless they were clean, fresh and neatly packed. Irish leaders of the co-operative movement responded to this by setting up the first poultry societies, and by 1900, 21 societies had been set up. However, co-operators did not believe that small societies such as these would provide a solution to the large-scale problem.[23] One of the difficulties identified by Horace Plunkett was the apparent reluctance of farming women to commit themselves to co-operative ideals. 'I am afraid she is not a good co-operator!'[24] As with co-operative creameries, however, many women saw the organization of the egg trade by the co-operatives as a threat to their income, as the men received the payment for eggs delivered to creameries along with the milk.

Poultry production was significantly helped by the Irish Department of Agriculture. The value of eggs exported to Britain rose between 1900 and 1911, so that Ireland had become its biggest supplier. In 1914, poultry accounted for 9% of Irish agricultural output, compared to 5–6% in the 1890s. All counties, apart from Antrim, set up hatcheries to supply farmers with high-quality, disease free hens. Instructresses trained in the Munster Institute, advised farm women on care of poultry. The Department of Agriculture, along with the co-operative movement, worked to ensure that eggs exported from Ireland to Britain, arrived fresh and clean. However, as late as 1911, the Department was warning egg producers, and egg merchants against holding on to eggs for too long.[25]

During the First World War, the Department of Agriculture warned Irish farmers that they were losing a golden opportunity in capturing the British egg market, as Britain could no longer import eggs from the continent. However,

124 Mrs Dora McArdle feeding hens at Mullaghduff, Cullyhanna,
County Armagh, photo courtesy of M. McSparran.

there were ongoing problems in getting eggs to Britain, including a shortage
of wood for packing cases, and less reliable shipping services.[26] The post-war
period was very difficult for egg producers. In 1921, the turn-over from co-
operative poultry societies dropped by 30%.[27] The southern government acted
to raise the standards of Irish eggs by passing the Agricultural Production
(Eggs) Act in 1924, which laid down standards for grading and packing export
eggs. This helped the reputation of Irish eggs abroad, but co-operative societies
formed to market poultry products continued to decline in numbers.

WOMEN AND POULTRY

The most significant aspect of the management of poultry was that during the
period covered by this book, it was almost entirely the responsibility of farm
women (fig. 124).[28] The notion that looking after chickens was women's work
was well-entrenched even during the 1950s. Rosemary Harris, writing about
farm people in County Tyrone, summarized the attitudes of men.

Looking after poultry and turkeys was women's work essentially, and neither the farmer nor his labourers would have dreamt of giving a hand in it. Once [the farmer's wife] … did ask … the hired man to help her and he was furiously indignant. Locally any male over about thirteen or fourteen would have considered himself insulted by such a request.[29]

However, women earned significant amounts of money from the egg trade, often selling them to egg traders, or local grocers. In the latter case, the eggs were often bartered for goods that could not be produced at home, such as tea and sugar. Farm women developed networks of mutual help in poultry production that paralleled the ties that bound many male farmers. Mrs Ellen Gibson, described how these operated around Lisbane in north County Down in the 1940s.

If you saw a neighbour's hens out early in the morning, well [you knew they] were good hens … good laying hens. And you went along to see if the neighbour would gather you two settings of eggs [for hatching] … and you would ask if she wanted them swapped, or did she want money for them … [In return for the swapped eggs] you had to take your eggs [for eating], and you had to give her a day or two to pick them, because you didn't want a great big egg … because sometimes they would have died in the shell when they were coming out … Sometimes in a big egg there was two eggs in one – a double yoked egg.

Women also helped one another in breeding turkeys.

Whatever farmer's wife had a turkey rooster, you had to take your [turkey] hen there … When she was sitting, you got her below your arm, and you took her … across the fields. You had to mark her, for maybe this woman had a dozen [turkey] hens. And you let her stay a day or two. Then you had to go back and bring her home, and if she didn't lay inside a fortnight, you had to take her back again.[30]

During the early twentieth century, there was a move towards more intensive methods of poultry production. One development was a working relationship between farm women, and local hatcheries. In County Fermanagh in the 1930s, for example, Mrs Mary Wilson kept a flock of about 150 Leghorn,

Wyandot and Sussex hens. However, after the Second World War she bought a brooder, a heated device under which newly hatched chicks could be reared. The brooder, together with day-old chicks, was supplied by Farbairn's hatcheries in Portadown. This was a move towards the large scale, specialized production of chickens and eggs, which became increasingly widespread during the later twentieth century. However, in the 1950s, Mrs Wilson continued to deal with the local grocer, selling him eggs once a week, and using the money earned to buy tea, sugar, bread and jam.[31] Leaders of the co-operative movement observed a similar reversion throughout Ireland during the Depression years of the 1930s, where eggs were used by women as a 'depreciated currency'.[32] From the 1950s, the increasing movement towards deep litter and battery farming of poultry, meant that women's role in managing production became less central.

CHAPTER EIGHTEEN

The Theory and Practice of Improvement

THE HISTORY OF Irish farming in the twentieth century was at least as complex as that of the previous centuries. Some very broad, and heavily qualified generalizations can be made. The long-term swing away from tillage to pastoral farming continued, with some dramatic short-term reversals during the century's two World Wars. These wars also entrenched the direct engagement of government with farming, including the protection of farmers' incomes through guaranteed prices. There was also ongoing mechanization of farming.

World War I brought a sharp rise in agricultural prices, compulsory tillage and the introduction of statutory wages for agricultural labourers, although by 1920, farmers had to adjust to falling prices.[1] The War of Independence which led to the formation of the Irish Free State and Northern Ireland through partition, and economic war with Britain, did not have the huge long-term effects on farming that might be expected. There was continuity in the work of the Department of Agriculture in the new Free State.[2] In both jurisdictions within Ireland, increased government intervention in farming was signalled by the Livestock Breeding Act in 1922 and the Marketing of Eggs Act of 1924 (Northern Ireland), and the Dairy Produce Act and the Agricultural Produce (Eggs) Act of 1924 in the Irish Free State.[3]

From the 1930s the market for Irish agricultural produce was regulated by price controls and other mechanisms. After 1932 the emphasis in the relationship between the Department of Agriculture and the farmers changed to one of price stabilization, bounties and compensation payments, subsidies to domestic producers, and measures designed to achieve the Irish government's self-sufficiency programme.[4] Similar developments occurred in Northern Ireland, where successful protection schemes were put in place, and state marketing schemes led to the huge growth in the number of pigs and exports of fat cattle.[5]

Within farming practice, there was increasing specialization, and mechanization. Ireland has a key place in the history of tractors, perhaps the most important single engineering development for farming in the first half of the century. In March 1917, the Department of Agriculture and Technical Instruction established a motor tractor section. At this time there were only 70

125 Fordson tractor, *c.*1915 (UFTM, WAG 1982).

tractors in Ireland; by September 1918, there were 640. The Department organised imports of tractors, supplies of spare parts, and tractor ploughs. Two-week training courses for tractor drivers were held at the Albert Agricultural College in Glasnevin. In 1918 and 1919, the Department was asked to investigate which types of tractor were suitable for Irish conditions, to organise exhibitions of farm machinery and to encourage the use of milking machines.[6]

Henry Ford, some of whose family came from Ireland, began to manufacture his Fordson tractors in 1917 (fig. 125). These were the first tractors to be produced on a massive scale, and at a price affordable by ordinary farmers. Fordson Model F tractors were produced in Cork in 1918, as part of Ford's support for the British war effort. Production slumped in the 1920s, and the Ford factory in Cork was closed in 1928, but it was re-opened in 1929, in response to the demand for tractors from the Soviet Union. The factory was transferred to Dagenham in England in 1932.[7]

In the north, Harry Ferguson, a County Down man, began work on his world famous Ferguson system in 1917. The system was revolutionary in that it transformed tractors from being simply haulage vehicles, to machines whose driver could also manipulate implements fixed behind them – raising and lowering, or moving them sideways, for example. This was achieved by a three-

126 Ferguson tractor harvesting oats, photo courtesy of the
Museum of English Rural life.

point linkage and hydraulic lift fitted to the back of the tractor. Ferguson-
Brown tractors, which incorporated the system, were manufactured in
Huddersfield in the David Brown factory in 1936. The 1936 prototype allowed
farmers to control basic tillage tools, such as ploughs and cultivators. Over the

next twelve years, the system was developed so that many more of the heavy tasks on the farm could be carried out using a tractor, including manure spreading, grass mowing, reaping grain, threshing, and transport of produce (fig. 126).

Ferguson and Ford met in the USA, shortly after the manufacture of Ferguson-Browns had started. The two men agreed that Ferguson's system could be incorporated in the new Ford tractors, the Ford 9N, with Ferguson having control over marketing the machines. This became known as the 'handshake agreement', and mass production of 9N tractors in the USA began. However, after Ford died in 1948, the agreement collapsed, and Ford's grandson Henry Ford II tried to take control of the marketing of the 9N from Ferguson. The ensuing court case, won by Ferguson, was one of the lengthiest and most expensive in history. Ferguson tractors were developed into the 1950s, and the TO35, 'The Wee Grey Fergie,' which was designed under the supervision of Herman Klemm, became known worldwide, with many still in use today.[8]

The Second World War led to even greater government intervention in farm production. Compulsory tillage was accompanied by guaranteed prices for agricultural prices, and grants allowed the accelerated mechanization of farming. In Northern Ireland, the spread of tractors was again a good indicator of this. At the start of the War, in 1939, there were only 550 tractors in Northern Ireland. By 1945, there were 7,300.[9] The expectation that government would support farming through grants and subsidies was well established by the end of the war, in both parts of Ireland, and indeed throughout Europe. One massive consequence of this was the Common Agricultural Policy of the European Common Market, the dismantling of which has proved a lengthy and controversial task.

GENERAL CONCLUSIONS

The old notion that Irish agriculture was backward and inefficient at any time during the last three hundred years is now largely discredited. By the early nineteenth century Irish systems of agriculture could stand comparison with Britain and north-west Europe.[10] When we consider contemporary criticisms of Irish farming practices in the light of how implements and techniques were actually used, we can often demonstrate the rationality of Irish farming methods.

During the later eighteenth and early nineteenth centuries, almost every general account of Irish cultivation techniques emphasized their defects. Common complaints were that rotations were too short, crops were weed-

infested, and that the land was often cropped to exhaustion. 'Common' agricultural implements were also dismissed as badly constructed and inefficient.
Tighe's general assertion of the state of Irish farming, published in 1802, is
typical:

> that mode of cultivation is, indeed, too well known in Ireland, by which
> a soil naturally fertile, but exhausted by repeated crops of corn, is
> abandoned to noxious weeds for several succeeding years: is again broken
> up, slightly manured, exhausted, and again abandoned: where culture,
> instead of improving, deteriorates; where no effort is made for perma-
> nent utility … and where the different branches of rural economy, so far
> from assisting each other, remain unconnected and distinct, in a state of
> natural repulsion.[11]

In their criticisms, agriculturalists tended to contrast the backwardness of Irish
cultivation techniques with those currently advocated, and to some extent
implemented, on English and Scottish farms. This 'improved' farming was
believed to arise from the systematic application of scientific ideas. There was
some recognition that, as with any successful experimentation, the theory and
practice of agriculture should enrich one another, but it was very rarely
conceded that Irish farming was also advanced by an interaction between
'improved' and 'common' farming methods.

IMPROVED PRACTICES

A very large number of farm implements and techniques were either introduced to or developed in Ireland during the last three centuries. However, it is
very difficult to assess the rate at which these came into widespread use. We
have seen that the Royal Dublin Society was importing small numbers of new
implements even before 1750, and some small-scale manufacturing of machines
and tools began in several parts of Ireland during the later part of the eighteenth century. All this activity seems to have accelerated during the early
nineteenth century. However, this is not necessarily evidence that most Irish
farmers were using new technology. One interesting, although very tiny survey
carried out by the Royal Dublin Society in 1859 suggests that adoption of new
machinery was not common even among fairly affluent farmers. Fourteen
farmers from ten different counties were questioned about their farms, labour

supply, and cultivation techniques. The farms varied considerably in size. The largest, in County Monaghan had 1,964 acres under tillage, while one of the County Antrim farms had just over five acres of cultivated land. All of the farms in the survey made some use of horse-power, with the number of working horses varying between one and six. Only two of the farms had reaping machines, which is not very surprising, since commercially viable reapers had been introduced to Ireland just seven years before. However, horse-drawn hay-tedding machines, which had been available much longer were also found on only two farms, and only five had horse-powered threshing machines. (Returns for two of the other farms give prices for the hire of a steam thresher.)[12]

The lack of use of these implements, which were proven to have great value in increasing both the possible scale and speed at which operations were performed, requires some explanation. Despite the enthusiasm of agriculturalists, a generally cautious attitude to new technology was well justified. Only a very small proportion of the implements and techniques devised by improvers were in fact successful. Large capital inputs and the use of advanced engineering were no guarantees that experiments would lead to practical advances in technique. This can be clearly seen in attempts to apply steam power to ploughing. Steam ploughing equipment was advertised in the *Irish Farmer's Gazette* in 1863, and exhibited at the Royal Dublin Society's Spring Show in 1864.[13] In 1863 *Purdon's Practical Farmer* was enthusiastic: 'Seeing that not only the possibility, but the actually greater utility of steam cultivation has been sufficiently proved, we have not the slightest hesitation in saying that the sooner it becomes a regular part of our agricultural system the better'.[14] However, despite the continued manufacture of steam ploughing apparatus, it appears to have been very little used in Ireland, and by the early twentieth century, farming texts had become much more cautious in their assessment of steam power:

> As an adjunct to horse cultivation, steam is extremely valuable, and in many cases, it may partially displace horses. It is, however, noteworthy that after a trial of at least sixty years, it has hitherto failed to oust the horse from its position as the most generally employed power on the farm … The efficiency of steam cultivation does not … appear to be in all cases greater than that of horse tillage. Horse ploughing is quite as good as steam ploughing, and, as to ordinary harrowing, it is one of its greatest recommendations on many soils that the horses tread the ground and render it firm after sowing.[15]

Just how long, and how inconclusive, a debate between agricultural improvers could be is well illustrated by arguments about the relative efficiency of oxen and horses as draught animals, discussed earlier. At a 'theoretical' level the debate never reached a satisfactory conclusion but during the nineteenth century, at the level of practice, the use of oxen dwindled away almost entirely.

The differing and constantly changing views of theoretical agriculturalists were often criticized by contemporary observers as hindering rather than advancing the cause of improvement amongst farmers. E. Burroughs, commenting on the progress of agricultural improvement in Ireland, complained: 'We find so much difference of opinion, even among the most experienced writers, that the young practitioner is often puzzled to decide on which he is to place confidence'.[16] This confusion also showed that the promotion of some improved practices and the rejection of some common practices by theorists were based on incomplete scientific knowledge. The common claim that trenching, which involved the inversion of the soil strata in deep tillage, enhanced soil fertility, and the rejection of paring and burning on the grounds as being an exhaustive process when it was used in reclamation of mountains and bogs, aptly illustrate this lack of knowledge.

THE PERSISTENCE OF 'COMMON' PRACTICES

Throughout this book we have seen that within limited capital resources, and where labour was plentiful and cheap, the 'common' practices generally dismissed by improvers could be efficient. Hand-tools might be simple in construction, but the techniques in which they were used were remarkable for refinement rather than crudity. Spadework and associated techniques of ridge cultivation show this very clearly, as does the use of sickles for harvesting grain. Indeed where there was so much under-employed labour, as in early nineteenth-century Ireland, we have seen that some agriculturalists, as well as socially concerned contemporaries, advocated that labour-intensive methods, such as using flails for threshing, or spadework for large-scale tillage, should be used as much as possible. The low level of mechanization of the farms in the Royal Dublin Society's survey mentioned earlier in this chapter was probably at least partly due to the availability of labour. All of the farms included had hired labour – one County Antrim farm employed 140 men and 120 women as extra hands during harvest.

The puzzlement of some agriculturalists when discussing common practices which still seemed to produce high yields is well illustrated by McParlan's comments on County Mayo in 1802:

> With the articles of implements I cannot have done without observing, that in general all through the county they are of the most inferior kind. The ploughs never sink deep enough, and the slight scratching they give the surface is left in unbroken lumps, by the light and wood-pinned harrows ... the rakes, shovels, and all are bad. It has appeared, notwithstanding, how immense is the extent of tillage here, and the vast abundance exported of potatoes and corn.[17]

Even where new types of implement had striking advantages in the speed and scale at which they could operate, or savings in draught they allowed, and where capital and labour were not limiting factors, a slow change to the new technology was by no means an indication of conservative prejudice. We have seen that 'common' Irish ploughs, for example, while lacking the refined construction of improved Scottish and English models, could be used efficiently in particular systems of cultivation. Early two-horse improved Scottish swing ploughs could only work to full advantage when ground was well drained, and where techniques such as the ploughing in of seed were not practised. An infrastructure of blacksmiths and ploughmen well versed in the principles of the new, improved plough technology was also necessary before the widespread manufacture and use of the implements was possible. This was lacking in late eighteenth-and early nineteenth-century Ireland, and as the development of such a tier of trained craftsmen and operatives takes time, it is not surprising that many farmers did not feel any urgent need to adopt the Scottish ploughs. New techniques, such as drilling grain seed, required a whole series of changes from tillage to harvest. The best agriculturalists recognised that a change in one aspect of a farming method could imply radical changes throughout the whole mode of cultivation.

Perceptive agriculturalists were well aware that systems of farming developed according to abstract principles, or to suit the circumstances of farmers in lowland Scotland or southern England, would often require modification before they would work well in Irish conditions. Arthur Young recorded one disastrous attempt to introduce English methods to an estate in south-west Ireland. The landlord, wishing to improve an area of very rocky mountain, brought over two farmers from Sussex. He had houses and out-buildings erected for them, and let them have the land rent-free. The men sold up their holdings in England, and brought over their horses, stock and implements. The land, however, proved too stony to be cleared at anything other than very great expense. The project continued for four years, but ended with the landlord losing money, and the two farmers ruined. Young saw that such an imprudent undertaking 'was the sure way to bring … ridicule on what falsely acquired the name of English husbandry'.[18]

The cultivation techniques described in this book have shown that Irish farmers cannot fairly be accused of stubborn intransigence. We can now re-examine some of the major inadequacies of Irish farming practice identified by contemporaries, and try and decide how far these defects resulted from deficiencies in farming at a technical level.

Minimal rotations and cropping to exhaustion were not inevitable results of Irish farming practice. The large proportion of arable land under grass, the frequent breaking up of pastures, and the natural fertility of soil in much of Ireland meant that good land, which remained largely under the control of commercial farmers, did not become exhausted. The practice of 'bare' or 'summer' fallowing, where land was repeatedly ploughed during the year, also maintained the fertility of land. Agriculturalists criticized this type of fallowing, advocating the use of green cropping instead, but most criticisms were directed against the waste of land involved, rather than the effectiveness of the practice. (In 1802, Tighe claimed that summer fallowing could only be economically practised in County Kilkenny on farms of more than fifty acres.)[19]

In areas of poorer land, where much of the expanding population in the late eighteenth and early nineteenth centuries was concentrated, subdivision of holdings meant that the 'common' controls on soil deterioration could not be applied. It was in these areas that rotations were shortened to become a 'see-saw between oats and potatoes, with occasional rests'[20] and where paring and burning might be carried on until land was ruined. Hely Dutton's comment on the latter practice in County Clare in 1808, shows the alternatives facing many small farmers: 'If a total abolition of this practice was to take place, as some people totally ignorant of rural economy seem to wish, a famine would be the consequence.'[21]

The claim that crops in Ireland were weed-infested can be related to the technique of broadcast-sowing cereals, and to late harvesting. Where crops were not sown in drills, weeding became extremely difficult, and the longer a crop was left standing, the more weeds grew amongst it. Reaping grain with sickles or hooks meant that weeds could be left behind during harvest, but how far weeds damaged crop yields remains uncertain. Conclusions are based on impressionistic evidence, since types of weeds, and their real extent, are not fully known. Agriculturalists differed in the emphasis they put on this problem. In 1835, it was alleged that a traveller in Ireland would see 'weeds often growing more luxuriant and numerous than even the grass or crops'.[22] However, this must be set against Arthur Young's comparison of the carefully weeded potato crops of the Palatines, in their colony at Adare, County Tipperary, with those of surrounding farmers: '[The Palatines] are different from the Irish in several particulars; they put their potatoes in with the plough, in drills, horse-hoe them while growing, and plough them out … but the crops are not so large as in the common method'.[23] Careful husbandry was essential if the related set of drill implements were to be used, but Young, and the Irish, thought the Palatines went too far: 'Their industry goes so far, that jocular reports of its excess are spread: in a very pinching season, one of them yoked his wife against a horse, and went in that manner to work, and finished a journey at plough.'[24]

One reason why agriculturalists disapproved of Irish methods seems to have been their untidy appearance. But this may have belied their effectiveness: a point which was recognized, not only by Arthur Young, but also by Hely Dutton who, as we have seen, warned landlords organizing the drainage of land not to mistake neatness of execution for correctness of design. Horatio Townsend, discussing the practice of paring and burning in County Cork in 1810, suggested that the naked appearance of land after burning was the main reason for objections to the technique.[25] John Sproule, writing on the effects of cross ploughing during 'summer' or 'naked' fallowing, noted that the plough cut through the previously ploughed ridges in such a way that the furrows were 'generally thrown up in an irregular manner, and present to the inexperienced eye a confused aggregation of lumps of earth'. However, he claimed that despite the untidy appearance the technique was 'well adapted for exposing a very enlarged surface of soil to the ameliorating influence of the atmosphere', and served as a check on the growth of weeds[26]

Appearance and notions of appropriate separation also underlay many of the criticisms of the management of livestock on small farms. The collection of dung in byre dwellings, and the housing of pigs and people in the same living space, produced reactions of disgust and derision, but inside the resources available to the poorest farmers, they were important ways in which supplies of fertilizer, meat, and rent money could be assured.

It would of course be absurd to present Irish farmers as uniformly good husbandmen. There were many recorded practices which can be criticized as inefficient or slovenly since, even within 'common' practice, better alternatives were available. Probably the most notorious example of bad husbandry was ploughing by the tail, where draught horses' tails were substituted for trace ropes. Justifications of this practice, that it shortened the overall length of the plough team, or that it made horses more sensitive to obstructions met by the plough, are not convincing.[27] Alleged instances where the horses' tails were pulled off during ploughing would mean that, as harness, tails were not very reliable, and certainly very difficult to repair. We can dismiss ploughing by the tail as bad husbandry because an efficient and widely used system of harnessing was within the resources of even small farmers. Horse collars made of plaited *súgán* (straw rope), and traces made from straw or withies, were recognised by contemporaries as easily made, cheap, and effective.

The fact that that Irish farming methods were generally efficient within the resources available, does not mean that agricultural research was in any way superfluous. The practical successes of this research have been as impressive as those of any other applied science. In 1863, *Purdon's Practical Farmer* admirably described the state of agricultural research:

> We have yet much to learn, and … in general, we are only groping after, and have yet but faint conceptions of what are the true principles by which we should be guided, or the correct modes of reducing those principles to practice. But the knowledge of our deficiencies on these points ought to induce us to strain every nerve in order that all doubts may be cleared away, and that agriculture, the most important in its results of all the arts, shall be placed in its proper station, based upon and guided by the truths of general science.[28]

In any situation, however, there is a lag between the introduction of the discoveries arising from these worthy aims and their widespread adoption.[29] Even in Norfolk, identified as one of the heartlands of the English agricultural revolution, the adoption of the famous four-course rotation was very slow. Joan Thirsk, discussing changes in English agricultural techniques, reaches a conclusion very similar to that suggested by Irish evidence: 'This survey of innovations underlines their hazards and uncertainties, which should not be brushed aside by afterknowledge. The risks of experiments were daunting to the great majority of farmers, who could not afford to gamble and lose'.[30] Contemporaries claimed that the lag between introduction and adoption, or at another level theory and practice, was wider in Ireland than in England and Scotland. We hope to have shown that at a technical level this lag did not imply inefficient farming. The history of flax, the most commercialized and least mechanized crop, shows very clearly how complex the relationship between technological change and economic development can be. Some integration of 'improved' and 'common' techniques was possible. Progressive, large-scale Irish farmers sometimes integrated elements from both systems, disregarding the more rigid strictures of theorists. One clear example of this was the use of high, narrow cultivation ridges in the initial stages of thorough drainage schemes, especially where marginal land was being reclaimed for cultivation. These reclamation schemes also led to the recognition that controlled paring and burning could accelerate the improvement of land. The integration of 'improved' and 'common' techniques was also practised by small farmers. This is especially clear when we look at the history of livestock farming. The most successful local breeds of horses, sheep, pigs and cattle (even Kerry cattle) were produced by breeding programmes which selected the best characteristics of local types of animal, and improved imports.

The history of tractor technology in Ireland shows the mixture of curiosity, caution and ingenuity with which twentieth-century farmers approached the potential of tractor technology. The owner of a small foundry near Rathfriland, County Down, remembered the caution of local farmers in the 1930s, very clearly. 'I mind Ferguson and he had all his equipment in the Belfast Show ... and he was ploughing and making drills and everything, and there was hardly anybody stopping to look at it!' The changing attitudes to tractors can be seen from the minutes of the Northern Ireland Ploughing Association (NIPA). In 1939, the committee of the Association agreed that tractors could be included in the International Match that year. As late as 1943, however, one member of the association stated that he was 'not in favour of tractors at all ... [It was difficult enough to get] a field big enough for horses, let alone tractors as well.' By 1945, however, the Association's secretary reported that there might be more prizes for tractors than horses at the match planned for that year.[31] The ingenuity of farmers in adapting the new tractor technology to their needs can

127 James Grimes with the double mouldboard plough fitting which he built for making ridges at Carrigallen Ploughing match in 2007.

be seen in their use in making cultivation ridges. As we have seen, lazy bed ridges were still being made in County Kerry in the 1970s, using a swing plough made for use with horses, pulled behind a tractor (fig. 50). The annual ploughing competition at Carrigallen, County Leitrim still has a section for ridges made in this way, and at least one local farmer is experimenting with modifications to ploughs to allow ridgemaking using tractors to become even more efficient (fig. 127).

The technical history of Irish agriculture is full of examples of the ingenuity with which farmers adopted or modified new techniques, if convinced of their value. Irish farmers were, however, very ready to ignore those which they did not regard as useful. As Peter Solar points out, 'There were elements of the faith [Improvement], notably the use of improved implements and machinery, which were probably inappropriate to Irish labour supply conditions. All in all Irish agriculturalists were eclectics. They resisted repeated entreaties to conform and so earned the condemnation of the true believers.'[32] The success of their approach was demonstrated by the high yields produced by Irish commercial farming, and the skill with which the smallest subsistence farmers created a living out of previously waste land.

Examining the relationship between the resources available to small farmers, and their farming techniques, raises another point that has been well made by Joel Mokyr. Although 'in a certain way ... the Irish economy was well organised and [the Irish] were not throwing away resources ... [This] is not quite the same as saying they were not poor'.[33] Farming techniques cannot be considered in isolation. The skills of ridgemaking and reaping developed by the Irish rural poor could not save vast numbers of them from extinction when the odds became overwhelming. The ongoing struggle for farming to survive in the new millennium, happily carried on in a much more affluent context, involves the same mixture of environmental, economic and social factors, but on a global scale. It remains to be seen whether farmers, the great class of survivors, will be able to overcome these obstacles raised in the future.

Notes

CHAPTER ONE *Rural Society and Farming Methods*

1   John Feehan, *Farming in Ireland* (Dublin, 2003), pp 29–92.
2   Lindsay Proudfoot, 'Landlords', in S.J. Connolly (ed.), *The Oxford companion to Irish history* (Oxford, 1998), p. 297.
3   Cormac Ó Gráda, 'The investment behaviour of Irish landlords, 1830–75: some preliminary findings', *Economic History Review*, 23 (1975), and Cormac Ó Gráda, *Ireland: a new economic history, 1780–1939* (Oxford, 1994), p. 30.
4   Ibid., p. 29.   5 Ibid., p. 257.
6   Arthur Young, *A tour in Ireland* vol. 2, part 1 (Dublin, 1780), p. 457.
7   Ó Gráda, *Ireland: a new economic history*, p. 32.   8 Ibid.
9   Young, *Tour in Ireland*, vol. 2. part 1, pp 162–3.
10 Jonathan Bell and Mervyn Watson, *Farming in Ulster* (Belfast, 1988), p. 5.
11 David Fitzpatrick, 'The disappearance of the Irish agricultural labourer, 1841–1912', *Irish Economic and Social History*, 7 (Dublin, 1980), 68.
12 Margaret Morse, 'Conacre', in *The encyclopaedia of Ireland* ed. Brian Lalor (Dublin, 2003), p. 226
13 Joseph Lambert, *Observations on the rural affairs of Ireland* (Dublin, 1829), p. 195.
14 Kevin O'Neill, *Family and farm in pre-Famine Ireland* (Madison, WI, 1984), pp 104–5.
15 Young, vol. 2, part 1, p. 27.   16 *Family and farm*, p. 104
17 Clachan is a Scottish word adopted by Estyn Evans in his search for an appropriate term to describe a rundale cluster of housing. In Scotland, however, the term is used to describe a bigger settlement, with churches, shops and other services. The settlements in runrig systems are usually known in English as 'townships'.
18 Desmond McCourt, 'Infield and outfield in Ireland', *Economic History Review*, 7 (1954–5), 373.
19 Tom Yager, 'What was rundale and where did it come from?' *Béaloideas*, 70 (Dublin, 2002), 159 and 164.
20 R. Buchanan, 'Field systems in Ireland' in A.R.H. Baker and R.A. Butlin (eds), *Studies in the field systems of the British Isles* (Cambridge, 1973), p. 596.
21 Royal Commission on Congestion in Ireland. Second Appendix. *Digest of evidence* (HMSO: Dublin, 1908), p. 335.
22 James Anderson, 'Rundale, rural economy and agrarian revolution: Tirhugh 1715–1855', in William Nolan, et al. (eds), *Donegal: history and society* (Dublin, 1995) p. 448.
23 Jonathan Bell and Mervyn Watson, 'Clachan project: a report for the Glens of Antrim Historical Society' (unpublished), 2006.
24 McCourt, 'Infield and outfield', 377.
25 Desmond McCourt, 'The rundale system in Donegal, its distribution and decline', *Donegal Annual*, 3 (1955), 48.
26 Ibid.
27 Royal Commission on Congestion, *Digest of evidence*, pp 335, 336 and 337.
28 Ibid., p. 337.
29 Young, *Tour in Ireland* vol. 2, part 2 (Dublin, 1780), p. 39.
30 John Lanktree, PRONI D/2977/6/3.   31 John Lanktree, PRONI D/2977/6/4A.
32 Buchanan, 'Field systems in Ireland' p. 596.
33 Lord George Hill, *Facts from Gweedore* (5th ed.) intro. E.E. Evans (Belfast, 1971).
34 Ibid., p. 3.
35 Fred Coll and Jonathan Bell, 'An account of life at Machaire Gathlán (Magheragallan) County Donegal early this century', *Ulster Folklife*, 36 (Holywood, Co. Down, 1990), 83.

36 Hill, *Facts from Gweedore*, p. 17.
37 We are grateful to Róise Ní Bhaoil for this information.
38 Anderson, Rundale, rural economy and agrarian revolution, p. 468.
39 1841 census returns, quoted in Feehan, *Farming in Ireland*, p. 120.
40 Feehan, *Farming in Ireland*, pp 132–3.
41 Liam Downey, 'A Foresight of Rural Ireland' (Dublin, 2003).

CHAPTER TWO *Farm Houses and Farm Yards*

1 Alan Gailey, *Rural houses in the North of Ireland* (Edinburgh, 1984).
2 F.H.A. Aalen, 'Buildings', in Aalen et al. (eds), *Atlas of the Irish rural landscape* (Cork, 1997), pp 153–4.
3 Conrad Arensberg, *The Irish countryman* (Gloucester, MA, 1959), pp 23–4.
4 Henry Glassie, *Passing the time in Ballymenone* (Philadelphia, 1982), pp 327–31.
5 Arthur Young, *Tour in Ireland*, vol. 2, part 2, 'General observations' (Dublin, 1780), pp 35–6.
6 Jonathan Bell, 'Miserable hovels and substantial habitations: the housing of rural labourers in Ireland since the eighteenth century', *Folklife*, 34 (Reading, 1996), 45.
7 Young, *Tour in Ireland*, vol. 2, 'General observations', p. 29.
8 Robbie Hannan and Jonathan Bell, 'The *Bothóg*: a seasonal dwelling from County Donegal' in Trefor M. Owen (ed.), *From Corrib to Cultra: essays: in honour of Alan Gailey* (Belfast, 2000).
9 Bell, 'Miserable hovels and substantial habitations', p. 45.
10 Paul Scott, *A division of the spoils* (St Albans, 1975), p. 579.
11 Joel Mokyr, *Why Ireland starved* (London, 1987).
12 William Makepeace Thackeray, *Works*, vol. 18, *An Irish Sketchbook of 1842* (London, 1879), p. 130
13 Mokyr, *Why Ireland starved*, p. 197.
14 E.T. Craig, *An Irish commune: the history of Ralahine* (Dublin, 1919), p. 23.
15 Bell, 'Miserable hovels and substantial habitations', p. 49.
16 Robert Beatson, 'On cottages', *Irish Agricultural Magazine* (Dublin, 1793), 139.
17 [Devon Commission] *Digest of Evidence … into the state of the law and practice in respect to the occupation of land in Ireland*, vol. 2, part 1 (Dublin, 1847), p. 129.
18 Martin Doyle, *A cyclopaedia of husbandry* (London, 1844), p. 160.
19 Kevin O'Neill, *Family and farm in pre-Famine Ireland* (Madison, WI, 1984), p. 119.
20 *Agricultural classbook* (Dublin, 1856), p. 55.
21 Francis Joseph Bigger, 'Labourer's cottages for Ireland', *Independent Newspaper* (Dublin, 1907), p. 7.
22 Young, *Tour in Ireland*, vol. 2, 'General observations', p. 221.
23 Revd Plymseley, 'On cottages' in Arthur Young (ed.), *The annals of agriculture* (London, 1804), pp 447–8.
24 *Ulster farmer and mechanic*, vol. 1 (Belfast, 1825), p. 156.
25 Ibid., pp 277–8.
26 Royal Commission on Congestion, *Tenth Report*, Appendix (HMSO: Dublin, 1908), p. vi.
27 T.W. Moody, *Davitt and the Irish revolution, 1846–1882* (Oxford, 1981), pp 423–4
28 F.H.A. Aalen, 'Public housing in Ireland, 1880–1921', *Planning Perspectives*, 2 (1987), 289.
29 John W. Boyle, 'A marginal figure: the Irish rural labourer', in S. Clarke and J.S. Donnelly (eds), *Irish peasants: violence and political unrest, 1780–1914* (Manchester, 1983), p. 333.
30 Jonathan Bell, *Ulster farming families* (Belfast, 2005), p. 21. Paddy Smith, 'Agriculture' in Brian Lalor (ed.), *The encyclopedia of Ireland* (Dublin, 2003), p. 12.
31 John M. Mogey, *Rural life in Northern Ireland* (Oxford, 1947), p. 181.
32 Ibid., p. 35.   33 Ibid., p. 37.
34 Young, *Tour in Ireland*, vol. 2, 'General observations', p. 36.
35 John Sproule, *A treatise on agriculture* (Dublin, 1839), pp 104–5.
36 Ibid., p. 39.

37   Joseph Lambert, *Observations on the rural affairs of Ireland* (Dublin, 1829), pp 48 and 147–8.
38   Sproule, *Treatise on agriculture*, pp 94–5.
39   Lambert, *Observations on the rural affairs of Ireland*, pp 150–1.
40   *Stephen's Book of the Farm*, vol. 2 (5th ed. rev.) (Edinburgh, 1908), p. 277.

CHAPTER THREE   *Ridges and Drains*

1   [Devon Commission] *Digest of evidence … into the state of the law and practice in respect to the occupation of land in Ireland*, part 1 (Dublin, 1847), p. 571.
2   James McParlan, *Statistical survey of the county of Donegal* (Dublin, 1802), p. 99.
3   *North West of Ireland Society Magazine*, 2 (Derry, 1824), 53.
4   [Devon Commission], *Digest of evidence …*, p. 603.
5   Ibid., pp 587–8.
6   Mr and Mrs S.C. Hall, *Ireland: its scenery and character*, vol. 3 (London, 1841), p. 262.
7   Ibid., pp 263–4.
8   [Devon Commission], *Digest of evidence …*, p. 42.   9 Ibid.
10   Hall, *Ireland: its scenery and character*, p. 266.
11   Arthur Young, *A tour in Ireland*, vol. 1 (Dublin, 1780), p. 112.
12   Ibid., vol. 2, part 1, p. 117.
13   *Ulster farmer and mechanic* (Belfast, 1824–5), p. 309. John Dubordieu, *Statistical survey of the county of Antrim* (Dublin, 1812), p. 202.
14   W.S. Trench, *Realities of Irish life* (London, 1868), p. 98.
15   [Devon Commission], *Digest of evidence …*, p. 79.
16   M.L. Parry, 'A typology of cultivation ridges in Southern Scotland', *Tools and Tillage*, 3:1 (Copenhagen, 1976), p. 12.
17   W. Tighe, *Statistical observations relative to the county of Kilkenny* (Dublin, 1802), p. 184.
18   *The farmer's friend* (London, 1847), p. 77.
19   [Devon Commission], *Digest of evidence …*, p. 100.
20   Ibid., p. 92.
21   [Devon Commission], *Digest of evidence …*, p. 116.
22   Joseph Lambert, *Agricultural suggestions to the proprietors and peasantry of Ireland* (Dublin, 1845), p. 28.
23   *North West of Ireland Society Magazine*, 1 (Derry, 1823), 87.
24   Lambert, *Agricultural suggestions*, p. 28.
25   [Devon Commission], *Digest of evidence …*, pp 84–5.
26   Young, *Tour in Ireland*, vol. 1, pp 146–50.
27   Ibid., pp 369–81   28 Ibid., p. 138.   29 Ibid., p. 111.
30   Joseph Archer, *Statistical survey of the County Dublin* (Dublin, 1801), p. 73.
31   C. Coote, *Statistical survey of the county of Cavan* (Dublin, 1802), p. 120.
32   John McEvoy, *Statistical survey of the county of Tyrone* (Dublin, 1802), p. 115.
33   Hely Dutton, *Statistical and agricultural survey of the county of Galway* (Dublin, 1824), pp 169–70.
34   James Smith, *Remarks on thorough drainage and deep ploughing* (Stirling, 1844).
35   [Devon Commission], *Digest of evidence …*, pp 81–2; Dutton, *Statistical … survey of the county of Galway* p. 173.
36   *Purdon's practical farmer* (Dublin, 1863), pp 166–7.
37   Irish National Schools, *Introduction to practical farming*, vol. 2 (Dublin, 1898), p. 362.
38   *Purdon's practical farmer*, p. 175.
39   R.P. Wright (ed.), *The standard cyclopaedia of modern agriculture and rural economy*, vol. 4 (London, 1909), pp 191.
40   Ibid., p. 192.   41 Ibid., p. 217.
42   K.H. Connell, 'The colonization of waste land in Ireland, 1780–1845', *Economic History Review*, 3:1 (London, 1951), 70.

CHAPTER FOUR *Manures and Fertilizers*

1 A.T. Lucas, 'Paring and burning in Ireland: a preliminary survey' in Alan Gailey and A. Fenton (eds), *The spade in Northern and Atlantic Europe* (Belfast, 1970), pp 99 and 101.

2 Arthur Young, *A tour in Ireland*, vol. 2, 'General Observations' (Dublin, 1780), p. 69.

3 C. Coote, *Statistical survey of the county of Cavan* (Dublin, 1802), p. 114.

4 T.J. Rawson, *Statistical survey of the county of Kildare* (Dublin, 1807), p. 142.

5 H. Townsend, *Statistical survey of the county of Cork* (Dublin, 1810), pp 543–4.

6 R. Thompson, *Statistical survey of the county of Meath* (Dublin, 1802), pp 275–6; Townsend, *Statistical survey … of Cork.*, p. 124; John Dubordieu, *Statistical survey of the county of Down* (Dublin, 1802), pp 175–6.

7 [Devon Commission] *Digest of Evidence … into the state of the law and practice in respect to the occupation of land in Ireland*, part 1 (Dublin, 1847), pp 610–11; Mervyn Watson, 'Flachters: their construction and use on an Ulster peat bog', *Ulster Folklife*, 25 (Holywood, 1979).

8 E.E. Evans, *Irish folkways* (London, 1957), p. 147.

9 Lucas, 'Paring and burning in Ireland', p. 112.

10 *Ordnance Survey Memoirs*, Box 26 II and VI, 1835.

11 Young, *Tour in Ireland*, vol. 1, p. 112.

12 [Devon Commission], *Digest of Evidence … ,* pp 66–7.

13 Ibid., p. 690.

14 Fred Coll and Jonathan Bell, 'An account of life in Machaire Gathlán (Magheragallan) early this century', *Ulster Folklife*, 36 (Holywood, 1990), p. 82.

15 Young, *Tour in Ireland*, vol. 2, 'General observations', p. 68.

16 George Hill, *Facts from Gweedore*, (Belfast, 1971), p. 17.

17 W.L. Micks, *An account of the … Congested Districts Board for Ireland, from 1891 to 1923* (Dublin, 1925), p. 253.

18 *Instructions to owners applying for loans for farm buildings* (Dublin, 1856), p. 19.

19 Thomas Baldwin, *Introduction to practical farming* (Dublin, 1877), p. 6.

20 Evans, *Irish folkways*, p. 274.

21 Young, *Tour in Ireland*, vol. 1, p. 362.   22 Ibid., p. 347.

23 Asenath Nicholson, *The Bible in Ireland.* Quoted in Frank O'Connor (ed.), *A book of Ireland* (Glasgow, 1959), p. 136.

24 J. Binns, *Miseries and beauties of Ireland*, vol. 2 (London, 1837), pp 131–2.

25 Young, *Tour in Ireland*, vol. 1, p. 194.

26 'Regulations for gathering seaweed, Ardglass, Co. Down'. Public Records Office of Northern Ireland. Document T1O09/160.

27 Testimony of Mr Patrick Brogan, Hornhead (Ulster Folk and Transport Museum Tape, R79.36).

28 Jonathan Bell, *Ulster farming families* (Belfast, 2005), pp 36 and 63.

29 *Purdon's practical farmer*, p. 110.

30 Young, *Tour in Ireland*, vol. 2, 'General observations', p. 68.

31 Hely Dutton, *Statistical survey of the county of Clare* (Dublin, 1808), pp 157–8.

32 John Sproule, *A treatise on agriculture* (Dublin, 1839), p. 42.

33 R.M. Young, *Historical notes of old Belfast* (Belfast, 1896), p. 159.

34 Thomas H. Mason, *The islands of Ireland,* 2nd ed. (London, 1938), p. 6.

35 Young, *Tour in Ireland*, vol. 2, 'General observations', p. 68.

36 W. Tighe, *Statistical observations relative to the county of Kilkenny*, p. 443.

37 Hely Dutton, *A statistical and agricultural survey of the county of Galway* (Dublin, 1824), p. 180.

38 Young, *Tour in Ireland*, vol. 1, pp 140, 170 and 189.

39 Ibid., vol. 1, pp 147–8.

40 Tighe, *Statistical observations relative to … Kilkenny*, p. 440.

41 Martin Doyle, *A cyclopaedia of husbandry* (rev. ed.) W. Rham (London, 1844), pp 347–55.

42 C. Davies, 'Kilns for flax-drying and lime-burning', *Ulster Journal of Archeology*, 3rd Series, 1 (Belfast, 1938), p. 80.

43   Sproule, *Treatise on agriculture*, p. 37.
44   Young, *Tour in Ireland*, vol. 1, pp 9, 16, 34, 156.
45   Ibid., vol. 2, 'General observations', pp 67, 98.   46 Ibid., p. 68.
47   Coote, *Statistical survey of the county of Armagh*, pp 239–40.
48   Dubordieu, *Statistical survey of the county of Down*, p. 183
49   W. Greig, *General report on the Gosford Estates in County Armagh, 1821* intro. F.M.L. Thompson and D. Tierney (HMSO: Belfast, 1976), p. 110.
50   Peter Solar, 'Agricultural productivity and economic development in Ireland and Scotland in the early nineteenth century' in T.M. Devine and David Dickson (eds), *Ireland and Scotland, 1600–1850* (Edinburgh, 1983), p. 79.
51   *Purdon's practical farmer*, pp 114–15. For an authoritative account of the use of all manures and fertilizers, see James Collins *Quickening the earth* (Dublin, 2008).

CHAPTER FIVE   *Spades*

1   Alan Gailey, *Spade making in Ireland* (Holywood, Co. Down, 1982), pp v–vi.
2   Ibid., p. 33.
3   A.T. Lucas, ' "The Gowl Gob", an extinct spade from County Mayo, Ireland', *Tools and Tillage*, 3:4 (Copenhagen, 1979).
4   Alan Gailey, 'The typology of the Irish spade' in Alan Gailey and A. Fenton (eds), *The spade in Northern and Atlantic Europe* (Belfast, 1970), p. 39.
5   Mervyn Watson, 'Ploughs and spades: a nineteenth century debate, part one', *Ulster Folklife*, 46 (Holywood, Co. Down, 2000), 62.
6   Alan Gailey, 'Spade tillage in south-west Ulster and north Connaught', *Tools and Tillage*, 1:4 (Copenhagen, 1971).
7   Isaac Weld, *Statistical survey of the county of Roscommon* (Dublin, 1832), pp 657–9.
8   Gailey, 'The typology of the Irish spade', p. 39.
9   Thomas H. Mason, *The islands of Ireland* 2nd ed. (London, 1938), p. 58.
10   We are grateful to Mr Tom Casey of Crosskillmena, County Mayo for telling us this term.
11   Jonathan Bell and Mervyn Watson, *Irish farming: implements and techniques, 1750–1900* (Edinburgh, 1986), pp 50–3.
12   E.E. Evans, *Irish folkways* (London, 1957), p. 144.
13   W.R. Wilde, *Catalogue of the antiquities of the museum of the Royal Irish Academy* (Dublin, 1857), pp 206–7.
14   Hely Dutton, *Statistical survey of the county of Clare* (Dublin, 1808), pp 65–6.
15   Edmund Murphy, 'Spadework', *Irish Farmer's and Gardener's Magazine*, 1 (Dublin, 1834), 25.
16   James MacParlan, *Statistical survey of County Leitrim* (Dublin, 1802), p. 30.
17   Evans, *Irish folkways*, p. 134.
18   Thomas Baldwin, *Introduction to practical farming*, 23rd ed. (Dublin, 1893), p. 148.
19   Mervyn Watson, 'Spades versus ploughs: a nineteenth century debate: part two', *Ulster Folklife*, 49 (Holywood, 2004), 58–9.
20   Mervyn Watson, 'Spades versus ploughs: a nineteenth century debate: part one', *Ulster Folklife*, 46 (Holywood, 2000), 61.
21   Watson, 'Spades versus ploughs: part two', 56.
22   Watson, 'Spades versus ploughs: part one', 62.
23   Irish National Schools, *Introduction to practical farming*, vol. 2 (Dublin, 1898), pp 369–70.
24   *Irish Farmer's Gazette and Journal of Practical Agriculture*, 9 (Dublin, 1850), 91.
25   Ibid., 97.
26   Austin Bourke, *The visitation of God?: The potato and the Great Irish Famine*, ed. Jacqueline Hill and Cormac Ó Gráda (Dublin, 1993), p. 100.
27   Watson, 'Spades versus ploughs: part one', p. 55.
28   Alexander Yule, *Spade and manual labour, with low or cheap farming, a certain means of removing Irish distress; applicable to England and Scotland* (Dublin, 1850), pp 8–9.

29 *Farmer's Gazette and Journal of Practical Horticulture*, 7 (Dublin, 1848), 449.

30 Yule, *Spade and manual labour*, pp 21–2.

31 *Farmer's Gazette and Journal of Practical Horticulture*, 8 (Dublin, 1849), 141.

32 Ibid.

33 Watson, 'Spades versus ploughs: part two', p. 67.

CHAPTER SIX *Ploughs*

1 Martin Doyle, *A cyclopaedia of practical husbandry*, rev. ed. W. Rham (London, 1844), p. 437.

2 Jonathan Bell, 'Recent evidence for the use of Ard ploughs in Ireland', *Sinsear* (Dublin, 1983).

3 Jonathan Bell and Mervyn Watson, 'Cultivation ridges in County Longford', *Archaeology Ireland*, 19:4 (Dublin, 2005).

4 Mervyn Watson, 'Common Irish plough types and tillage practices', *Tools and Tillage*, 5:2 (Copenhagen, 1985).

5 Jonathan Bell and Mervyn Watson, *Irish farming: implements and techniques, 1750–1900* (Edinburgh, 1986), pp 65–7.

6 Ibid., pp 68–9.   7 Ibid.   8 Ibid., p. 70.

9 Thomas Hale, *A compleat body of husbandry*, 4 vols (Dublin, 1757).

10 John Wynn Baker, *A short description and list, with the prices of the instruments of husbandry, made in the factory at Laughlinstown, near Cellbridge, in the county of Kildare* (Dublin, 1767–9).

11 Arthur Young, *A tour in Ireland*, vol. 2 (Dublin, 1780), pp 32, 56, 123, 128.

12 Horatio Townsend, *Statistical survey of the county of Cork* (Dublin, 1810), p. 613.

13 G.E. Fussell, *The farmer's tools, 1500–1900* (London, 1952), p. 49.

14 Hely Dutton, *Observations on Mr. Archer's statistical survey of the county of Dublin* (Dublin, 1802), p. 33.

15 *Dublin Penny Journal*, 2 (Dublin, 1833), 150.

16 Richard Thompson, *Statistical survey of the county of Meath* (Dublin, 1802), p. 114.

17 Ibid.   18 Ibid., p. 113.

19 William Shaw Mason, *A statistical account and parochial survey of Ireland*, vol. 3 (Dublin, 1819), p. 633.

20 Hely Dutton, *Statistical survey of the county of Clare* (Dublin, 1808), p. 37.

21 *Munster Farmer's Magazine*, 1 (Cork, 1812), 85.

22 Ibid., p. 342.

23 *Munster Farmer's Magazine*, 2 (Cork, 1813), 85–7.

24 *Munster Farmer's Magazine*, 4 (Cork, 1816), 152–3.

25 Ibid., 78.

26 Jonathan Binns, *Miseries and beauties of Ireland*, vol. 2 (London, 1837), p. 296.

27 Bell and Watson, *Irish farming: implements and techniques, 1750–1900*, p. 77.

28 Mervyn Watson, 'North Antrim swing ploughs: their construction and use', *Ulster Folklife*, 28 (1982).

29 Jonathan Bell, 'The use of oxen on Irish farms since the eighteenth century', *Ulster Folklife*, 29 (1983).

30 A.T. Lucas, 'Irish ploughing practices, part 2', *Tools and Tillage*, 2 (Copenhagen, 1973), 78.

31 Jonathan Bell, 'Straw harness in Ireland', *Ulster Folklife*, 24 (1978). Doyle, *A cyclopaedia of practical husbandry*, p. 412.

32 C. Hyndman, *A new method of raising flax* (Belfast, 1774), p. 15.

33 W. Tighe, *Statistical observations relative to the county of Kilkenny* (Dublin, 1802), p. 201.

34 Doyle, *A cyclopaedia of practical husbandry*, p. 496.

35 Young, *Tour in Ireland*, vol. 1, pp 219a and 33.

36 Ibid., vol. 2, 'General observations', p. 225.

37 Jonathan Bell, 'A contribution to the study of cultivation ridges in Ireland', *Journal of the Royal Society of Antiquaries of Ireland*, 114 (Dublin, 1984), p. 81.

38 Caoimhín Ó Danachair, 'The spade in Ireland', *Béaloideas*, 31 (Dublin, 1963).
39 Tighe, *Statistical observations relative to the county of Kilkenny*, p. 295. Doyle, *A cyclopaedia of practical husbandry*, p. 496.
40 Ibid., p. 195.
41 Jonathan Bell and Mervyn Watson, 'A horse plough and harrows from County Leitrim' in Marion Meek (ed.), *The modern traveller to our past* (Belfast, 2006).

CHAPTER SEVEN   *Harrows, Rollers and Techniques of Sowing Seed*

1   Jonathan Bell, 'Harrows used in Ireland', *Tools and Tillage*, 4:4 (Copenhagen, 1983), 195.
2   Martin Doyle, *Hints to smallholders on planting and on cattle, etc.* (Dublin, 1830), p. 71.
3   R. Thompson, *Statistical survey of the county of Meath* (Dublin, 1802), p. 116; W. Tighe, *Statistical observations relative to the county of Kilkenny* (Dublin, 1802), p. 300. John McEvoy, *Statistical survey of the county of Tyrone* (Dublin, 1802), p. 47. Bell, 'Harrows used in Ireland', p. 197.
4   John Wynn Baker, *A short description and list, with the prices of the instruments of husbandry, made in the factory at Laughlinstown, near Cellbridge, in the county of Kildare* (Dublin, 1767–9), p. 30. Thomas Hale and others, *A compleat body of husbandry*, vol. 2 (Dublin, 1757), p. 105.
5   Tighe, *Statistical observations relative to the county of Kilkenny*, p. 500.
6   Hely Dutton, *Statistical survey of the county of Clare* (Dublin, 1808). p. 63.
7   Hale, *Compleat body of husbandry*, p. 104
8   Doyle, *Hints to smallholders*, p. 71.
9   Bell, 'Harrows used in Ireland', p. 200.
10 Baker, *Short description … of the instruments of husbandry*, p. 12.
11 Hely Dutton, *A statistical and agricultural survey of the county of Galway* (Dublin, 1824), p. 90.
12 Testimony of James Ennis, in Jonathan Bell, *Ulster farming families* (Belfast, 2005), p. 82.
13 Thompson, *Statistical survey of the county of Meath*, pp 116–17; Tighe, *Statistical observations relative to the county of Kilkenny*, p. 301.
14 C. Coote, *Statistical survey of the county of Armagh* (Dublin, 1804), p. 169.
15 E.E. Evans, *Irish heritage* (Dundalk, 1949), p. 91.
16 John O'Loan, 'A history of early Irish farming', part 2, Department of Agriculture (Ireland), *Journal*, 61 (Dublin, 1964), 19.
17 Bell, 'Harrows used in Ireland', p. 202.
18 Hale, *Compleat body of husbandry*, p. 112.
19 Thomas Baldwin, *Introduction to practical farming*, 23rd ed. (Dublin, 1893), pp 155–6
20 Martin Doyle, *Cyclopaedia of practical husbandry* (London, 1844), p. 497.
21 Baldwin, *Introduction to practical farming*, p. 157; John Sproule, *A treatise on agriculture* (Dublin, 1839), p. 701
22 Baldwin, *Introduction to practical farming*, p. 157.
23 Baker, *Short description … of the instruments of husbandry*, p. 29
24 *Munster Farmer's Magazine*, 2 (Cork, 1813), pp 137–41.
25 Doyle, *Cyclopaedia of practical husbandry*, p. 497; *Purdon's practical farmer*, p. 308; Baldwin, *Introduction to practical farming*, p. 156.
26 E. Wakefield, *An account of Ireland, statistical and politcal*, vol. 1 (London, 1812), p. 362; J. Binns, *The miseries and beauties of Ireland*, vol. 1 (London, 1837) pp 391–2.
27 Arthur Young, *A tour in Ireland*, vol. 2, part 1 (Dublin, 1780), p. 208.
28 Ibid., vol. 1, p. 72.
29 Baldwin, *Introduction to practical farming*, p. 15.
30 E.E. Evans, *Irish folkways* (London, 1957), pp 142–3.
31 Interview with Cormac McFadden, Roshin, Co. Donegal (Ulster Folk and Transport Museum Tape R80.39). Interview with John Joe McIlroy, Derrylea, Co. Monaghan (Ulster Folk and Transport Museum Tape R85.143).
32 Michael Partridge, *Farm tools through the ages* (Reading, 1973), p. 113; Sproule, *Treatise on agriculture*, p. 246.

33 Partridge, *Farm tools through the ages*, p. 119; *Stephens' Book of the Farm*, vol. 2, 5th ed. (Edinburgh, 1908), p. 124

34 Partridge, *Farm tools through the ages*, p. 120.

35 Evans, *Irish folkways*, p. 146.

36 Sproule, *Treatise on agriculture*, pp 71–3.

37 *Munster Farmer's Magazine*, 3 (Cork, 1814), 35.

38 Sproule, *Treatise on agriculture*, p. 247; John Sproule (ed.), *The Irish Industrial Exhibition of 1853; a detailed catalogue of its contents* (Dublin, 1854), p. 211; *Purdon's practical farmer*, p. 240.

39 *Munster Farmer's Magazine*, 3 (Cork, 1814), 133.

40 Sproule, *Treatise on agriculture*, pp 246–7

41 Doyle, *Cyclopaedia of practical husbandry*, pp 45 and 215.

42 Ibid., p. 569.

43 Tighe, *Statistical observations relative to the county of Kilkenny*, pp 200–1.

44 Thomas Baldwin, *Introduction to Irish farming* (London, 1874), p. 26.

CHAPTER EIGHT *Potatoes*

1 Joel Mokyr, *Why Ireland starved* (London, 1983); Cormac Ó Gráda, *Black '47 and beyond: the Great Irish Famine in history, economy and memory* (Princeton, 1999).

2 R.N. Salaman, *The history and social influence of the potato* (Cambridge, 1949), p. 222.

3 R.N. Salaman, 'The influence of the potato on the course of Irish history' (Tenth Finlay Memorial Lecture) (Dublin, 1944), p. 17.

4 Young, *A tour in Ireland*, vol. 1 (Dublin, 1780), p. 261, and vol. 2, 'General observations', pp 33–4.

5 Frank Mitchell, *The Irish landscape* (London, 1976), p. 205.

6 Young, *Tour in Ireland*, vol. 2, 'General observations', pp 33–4.

7 Salaman, *The history and social influence of the potato*, p. 288.

8 Ibid., pp 162–8.

9 Interview with Mr W. Donnelly (Ulster Folk and Transport Museum Tape C77.49).

10 Young, *Tour in Ireland*, vol. 2, part 1, p. 9 and pp 262–3.

11 De Latocnaye, *A Frenchman's walk through Ireland, 1796–7*, (trans. J. Stevenson) (Belfast, 1917), p. 123.

12 Young, *Tour in Ireland*, vol. 2, 'General observations', p. 222.

13 T.J. Rawson, *Statistical survey of the county of Kildare* (Dublin, 1807), p. 152; William Shaw Mason, *A statistical account, or parochial survey of Ireland*, vol. 1 (Dublin, 1814), p. 263; Young, *Tour in Ireland*, vol. 1, p. 85.

14 Alan Gailey, 'Spade tillage in south-west Ulster and north Connaught', *Tools and Tillage*, 1:4 (Copenhagen, 1971), pp 228–9; Jonathan Bell, 'A contribution to the study of cultivation ridges in Ireland', *Journal of the Royal Society of Antiquaries of Ireland*, 114 (Dublin, 1984), 88.

15 G.V. Sampson, *A memoir, explanatory of the chart and survey of the county of London-Derry, Ireland* (London, 1814), p. 304.

16 Interview with Mr Joe Kane, Drumkeeran, Ederney, County Fermanagh (Ulster Folk and Transport Museum tape R.84.40).

17 *Munster Farmer's Magazine*, 2 (Cork, 1813), 109; Young, *Tour in Ireland*, vol. 1, pp 33, 73, 79, 169, 209, and 215.

18 County Agricultural Instructors, *Agriculture in Ireland* (Dublin, 1907), pp 10 and 61.

19 John Wynn Baker, *A short description and list, with the prices of the instruments of husbandry, made in the factory at Laughlinstown, near Cellbridge in the county of Kildare* (Dublin, 1767), p. 7.

20 Young, *Tour in Ireland*, vol. 1, p. 211; vol. 2, part 1, p. 245.

21 Thomas Baldwin, *Introduction to practical farming*, 23rd ed. (Dublin, 1893), p. 11.

22 John Sproule, *A treatise on agriculture* (Dublin, 1839), p. 300.

23 John Wynn Baker, *Experiments in agriculture, made under the direction of the right honourable Dublin Society in the year 1767* (Dublin, 1769), p. ix.

24 W. Tighe, *Statistical observations relative to the county of Kilkenny* (Dublin, 1802), pp 220–1.

25 *Munster Farmer's Magazine*, 1 (Cork, 1812), 73; vol. 3, p. 253.

26 *Munster Farmer's Magazine*, 3 (Cork, 1814), 253.

27 Ibid., pp 52–3

28 Edmund Murphy, 'Agricultural report', *Irish Farmer's and Gardener's Magazine*, 1 (Dublin, 1834), 556

29 Jonathan Bell, 'Wooden ploughs from the Mountains of Mourne, Ireland', *Tools and Tillage*, 4:1 (Copenhagen, 1980).

30 R. Buchanan, 'Calendar Customs, part 1', *Ulster Folklife*, 8 (Belfast, 1962), 33.

31 Young, *Tour in Ireland*, vol. 1, p. 18; vol. 2, part 1, pp 9 and 90.

32 *Dublin Penny Journal*, 2 (Dublin, 1834), 283.

33 H. Townsend, *Statistical survey of the county of Cork* (Dublin, 1810), pp 656–7.

34 Rawson, *Statistical survey of the county of Kildare*, p. 153.

35 Patrick MacGill, *Children of the dead end* (London, 1914), p. 75; John McGahern, *The dark* (London, 1965).

36 *Patents for inventions: abridgements of specifications, class 6, agricultural appliances … A.D. 1855–6* (HMSO: London, 1905), p. 13.

37 Martin Doyle, *A cyclopaedia of agriculture* (rev. ed.) (London, 1844), p. 470; Sproule, *A treatise on agriculture*, p. 304; *Purdon's practical farmer*, p. 339.

38 Doyle, *Cyclopaedia of agriculture*, p. 470.

39 Murphy, 'Agricultural report', 556; Mason, *A statistical account …*, p. 260

40 Revd G.S. Cotter, 'Advice to the farmers of the county of Cork, on the failure of wheat', *Munster Farmer's Magazine,* 1 (Cork, 1812), 150.

41 Thomas P. O'Neill, 'The scientific investigation of the failure of the potato crop in Ireland, 1845–6', *Irish Historical Studies*, 5:18 (Dublin, 1946), 123.

42 Ibid., 125–6.   43 Ibid., 128

44 *Purdon's practical farmer*, p. 340.

45 O'Neill, 'The scientific investigation of the failure of the potato crop', 133–4.

46 W.L. Micks, *An account of … the Congested Districts Board for Ireland, from 1891 to 1923* (Dublin, 1925), p. 33.

47 Department of Agriculture and Technical Instruction for Ireland, 'Prevention of Potato Blight', *Journal*, vol. 2 (Dublin: HMSO, 1902), 684–90.

48 Cormac Ó Gráda, 'Economy, 1700–1922' in Brian Lalor (ed.), *The encyclopaedia of Ireland* (Dublin, 2003), p. 337.

49 Salaman, *History and social influence of the potato*, p. 321.

50 M.G. Wallace, 'Early Potato Growing', Department of Agriculture and Technical Instruction for Ireland, *Journal*, 6 (Dublin: HMSO, 1906), pp 3 and 9.

51 Salaman, *History and social influence of the potato*, p. 332.

CHAPTER NINE *Hay-making*

1 Bede, *A history of the English church and people* (trans. Leo Sherley-Price; rev. ed. R.E. Latham) (Penguin, 1983), p. 39.

2 G.B. Adams, 'Work and words for haymaking', *Ulster Folklife*, 13 (Holywood, Co. Down, 1967), 69.

3 [Devon Commission], *Digest of evidence taken before Her Majesty's Commissioners of Inquiry into the state of the law and practice in respect to the occupation of land in Ireland*, vol. 1 (Dublin, 1847), pp 635–6.

4 Arthur Young, *A tour in Ireland*, vol. 1 (Dublin, 1780) pp 517, 172.

5 Young, *Tour in Ireland*, vol. 2, 'General observations', pp 27 and 195.

6 Hely Dutton, *Observations on Mr Archer's statistical survey of the county of Dublin* (Dublin, 1802), p. 74.

7 R. Thompson, *Statistical survey of the county of Meath* (Dublin, 1802), p. 213; Thomas Baldwin, *Introduction to Irish farming* (London, 1874), p. 37.

8   Thompson, *Statistical survey of the county of Meath*, p. 213.
9   Young, *Tour in Ireland*, vol. 2, 'General observations', p. 139.
10 Information from Dr Austin O'Sullivan of the Irish Agricultural Museum, Wexford, and Mr James O'Kane of Claudy, Co. Derry.
11 Young, *Tour in Ireland*, vol. 2, part 1, pp 219–20, Hely Dutton, *A statistical and agricultural survey of the county of Galway* (Dublin, 1824), p. 136.
12 Department of Agriculture and Technical Instruction for Ireland, 'Haymaking', Leaflet no. 46, *Journal*, vol. 4 (Dublin: HMSO, 1904), p. 716.
13 *Irish Farmer's Gazette* (Dublin, 1863), pp 242.
14 *Purdon's practical farmer* (Dublin, 1863), pp 356–7.
15 Department of Agriculture and Technical Instruction for Ireland, 'Haymaking', p. 717.
16 Department of Agriculture and Technical Instruction for Ireland, *Twelfth annual report, 1911–1912* (London: HMSO, 1913), p. 45.
17 Young, *Tour in Ireland*, vol. 2, p. 246.   18 Ibid., p. 140
19 Dutton, *Observations on Mr Archer's statistical survey of the county of Dublin*, p. 74.
20 John Sproule, *A treatise on agriculture* (Dublin, 1839), pp 419–20.
21 *Cassell's household guide*, vol. 4 (London, 1880), p. 13.
22 *Purdon's practical farmer*, pp 359–61.
23 C. Coote, *A statistical survey of the county of Cavan* (Dublin, 1802), p.79, *Purdon's practical farmer*, pp 360–1; W. Tighe, *Statistical observations relative to the county of Kilkenny* (Dublin, 1802), pp 378–80; Wright, *Standard cyclopaedia of modern agriculture*, p. 240.
24 Sproule, *Treatise on agriculture*, p. 240, Thompson, *Statistical survey of the county of Meath*, p. 217.
25 Dutton, *Observations on Mr Archer's statistical survey of the county of Dublin*, p. 77.
26 *Farmer's Gazette*, 7 (Dublin, 1847), 137.
27 Sproule, *Treatise on agriculture*, p. 421, Martin Doyle, *A cyclopedia of husbandry* (rev. ed.) W. Rham (London, 1844). pp 280–1, Department of Agriculture and Technical Instruction for Ireland, 'Haymaking', p. 718; *Purdon's practical farmer*, p. 363.
28 Doyle, *Cyclopedia of husbandry*, p. 281
29 North West of Ireland Society, *Magazine*, 1 (Derry, 1823), 267.
30 R. Patrick Wright (ed.), *The standard cyclopaedia of modern agriculture and rural economy* (London, 1909), p. 245; *Stephens' Book of the Farm*, vol. 2, 5th ed. (Edinburgh, 1908), p. 277.
31 Baldwin, Thomas, *Introduction to Irish farming*, p. 42.
32 Hely Dutton, *Statistical survey of the county of Clare* (1808), p. 125.
33 Ibid., p. 126.
34 Wright, *Standard cyclopaedia of modern agriculture*, p. 240
35 Thompson, *Statistical survey of the county of Meath*, p. 215.
36 *Farmer's Gazette*, 7 (Dublin, 1847), 83–4.   37 Ibid.

CHAPTER TEN *Flax*

1   Jonathan Bell and Mervyn Watson, *Irish farming: implements and techniques, 1750–1900* (Edinburgh, 1986), pp 156–77. Arthur Young, *A tour in Ireland*, vol. 1 (Dublin, 1780), pp 59 and 141.
2   John Sproule, *A treatise on agriculture* (Dublin, 1839), p. 358.
3   A.I. Clark, 'Historical sketch of the flax-growing industry', Department of Agriculture and Technical Instruction for Ireland, *Journal*, vol. 3 (Dublin: HMSO, 1903), pp 688, 693, 696, 702; Henry F. Berry, *A history of the Royal Dublin Society* (London, 1915), p. 57.
4   R. Patrick Wright (ed.), *The standard cyclopedia of modern agriculture and rural economy*, vol. 6 (London, 1908), p. 12.
5   Clark, 'Historical sketch of the flax-growing industry', p. 689.
6   Ibid., p. 699.
7   W.H. Crawford, 'Economy and society in eighteenth-century Ulster', PhD thesis (Queen's University Belfast, 1982), pp 699–700.

8　C. Hyndman, *A new method of raising flax* (Belfast, 1774), p. 33. Thanks to Robbie Hannan for the standard Irish version of the saying.

9　Crawford, 'Economy and society in eighteenth century Ulster', p. 91.

10　Arthur Young, *Tour in Ireland,* vol. 1, pp 174–5.

11　Ibid., vol. 2, 'General observations', p. 155, Irish National Schools, *Introduction to practical farming,* vol. 1 (Dublin, 1898), p. 54; Wright, *Standard cyclopedia of modern agriculture,* vol. 6, p. 12.

12　Hyndman, *New method of raising flax,* pp 11–12.

13　Ibid., pp 13–14.

14　Arthur Young, *Tour in Ireland,* vol. 1, pp 163 and 183.

15　A.L. Clark, 'Flax seed for sowing purposes', Department of Agriculture and Technical Instruction for Ireland, *Journal,* vol. 4 (HMSO: Dublin, 1904), p. 269.

16　Wright, *Standard cyclopedia of modern agriculture,* vol. 6, p. 13.

17　Ulster Folk and Transport Museum, Specimen number 209.1983

18　*Munster Farmer's Magazine,* 5 (Cork, 1817), 148

19　Young, *Tour in Ireland,* vol. 1, p. 165.

20　Hyndman, *New method of raising flax,* p. 22; *North West of Ireland Society Magazine* 1 (Derry, 1823), 38. Wright, *Standard cyclopedia of modern agriculture,* vol. 6, p. 13.

21　Ulster Folk and Transport Museum tape R82.95.

22　James Silcock et al., 'The influence of rippling', Department of Agriculture and Technical Instruction for Ireland, *Journal,* vol. 3 (Dublin: HMSO, 1903), p. 680; Ulster Folk and Transport Museum tape R82.95.

23　Hyndman, *New method of raising flax,* p. 25.

24　Ulster Folk and Transport Museum tape R82.95.

25　*Munster Farmer's Magazine,* 4 (Cork, 1816), 255.

26　Young, *Tour in Ireland,* vol. 2, 'General observations', pp 162–3.

27　James McCully, *Letters by a farmer* (Belfast, 1787), p. 7

CHAPTER ELEVEN　*The Grain Harvest*

1　Jonathan Bell and Mervyn Watson, *Irish farming; implements and techniques, 1750–1900* (Edinburgh, 1986), pp 178–204.

2　Joseph Lambert, *Observations on the rural affairs of Ireland* (Dublin, 1829), p. x.

3　L.M. Cullen, *An economic history of Ireland since 1600* (London, 1972), p. 118.

4　Department of Agriculture and Technical Instruction for Ireland, 'Statistical Survey of Irish agriculture', *Ireland industrial and agricultural* (Dublin, 1902), p. 308.

5　John Feehan, *Farming in Ireland* (Dublin, 2003), pp 152–5

6　E. Burroughs, *The Irish farmer's calender* (Dublin, 1835), p. 216.

7　Thomas Baldwin, *Introduction to practical farming* (Dublin, 1877), p. 216.

8　*Belfast Newsletter,* 3 August 1750.

9　Arthur Young, *A tour in Ireland,* vol. 1 (Dublin, 1780), p. 161.

10　Martin Doyle, *A cyclopaedia of practical husbandry,* rev. ed. W. Rham (London, 1844), pp 44, 280; John Sproule, *A treatise on agriculture* (Dublin, 1839), p. 155.

11　Testimony of John O'Connor, Screen, Co. Wexford (Ulster Folk and Transport Museum tape R90.81)

12　'Pairlement Chloinne Tomáis', trans. O.J. Bergin, *Gadelica* (Dublin, 1912), p. 43.

13　W. Howatson, 'Grain harvesting and harvesters', in T.M. Devine (ed.), *Farm servants and labourers in lowland Scotland* (Edinburgh, 1984), p. 129.

14　J.C. London, *An encyclopaedia of agriculture* (London, 1831), p. 373.

15　Doyle, *Cyclopaedia of practical husbandry,* p. 493.

16　Francis Davis, 'Low and clean', in *Earlier and later leaves* (Belfast, 1878), p. 270.

17　A. Fenton, *Scottish country life* (Edinburgh, 1976), p. 54.

18　T.F. O'Rahilly, 'Etymological Notes, 2', *Scottish Gaelic studies,* 2 (Edinburgh, 1927), 26.

19 Young, *Tour in Ireland*, vol. 1, p. 72.
20 W. Tighe, *Statistical observations relative to the county of Kilkenny*, p. 214.
21 J.J. Murphy, 'The mower from Moygannon', *Ulster Folklife*, 12 (Holywood, 1966), 108.
22 J.A. Perkins, 'Harvest technology and labour supply in Lincolnshire and the East Riding of Yorkshire, 1750–1800', *Tools and Tillage*, 2:3 (Copenhagen, 1977), 128.
23 Amhlaoibh Ó Súileabháin, *Cinnlae Amhlaoibh Úi Shúileabháin*, vol. 3, trans. M. McGrath (Irish Texts Society: Dublin, 1930), p. 66.
24 Irish National Schools, *Agricultural classbook* (Dublin, 1868), p. 155
25 Patrick S. Dineen, *Foclóir Gaedhilge agus Béarla* (Irish Texts Society: Dublin, 1927), p. 1095.
26 Jonathan Bell, 'Sickles, hooks and scythes in Ireland', *Folklife*, 19 (Leeds, 1981), 31.
27 E.J.T. Collins, *Sickle to combine* (Reading, 1969), p. 9.
28 Tighe, *Statistical observations relative to the county of Kilkenny*, p. 184; Burroughs, *Irish farmer's calender*, p. 265.
29 Burroughs, *Irish farmer's calender*, p. 265.
30 *Irish Farmer's Gazette*, 29 August 1863 (Dublin), p. 297.
31 Doyle, *Cyclopaedia of practical husbandry*, p. 155.
32 G.E. Fussell, *The farmer's tools* (London, 1952), p. 116.
33 *Belfast Commercial Chronicle*, 1 October 1806.
34 Fussell, *Farmer's tools*, p. 129.
35 John Sproule (ed.), *The Irish Industrial Exhibition of 1853: a detailed catalogue of its contents* (Dublin, 1854), p. 216.
36 *Weekly Agricultural Review*, 15 April 1859, Supplement, p. viii.
37 *Irish Farmer's Gazette* 29 August 1863 (Dublin), pp 132–3 and 301. The *Impartial Reporter*, 3 August, 1967, claimed that a Pierce No. 1 reaping machine in Ballybay, Co. Monaghan, had originally been bought in 1864.
38 Sproule, *Irish Industrial Exhibition of 1853*, p. 215.
39 Advertisement for McCormick harvesting machines, *Dungannon News*, 1 May 1902.
40 Stephens, *Stephen's book of the farm*, 3 vols (5th ed.) (Edinburgh, 1908), p. 184.
41 Sproule, *The Irish Industrial Exhibition of 1853*, p. 250.
42 *Stephen's book of the farm*, p. 186.   43 *Purdon's practical farmer*, p. 284.

CHAPTER TWELVE *Threshing and Winnowing Grain*

1 Caoimhín Ó Danachair, 'The flail and other threshing methods', *Journal of the Cork Historical and Archaeological Society*, 60:19 (Cork, 1955).
2 E. Wakefield, *An account of Ireland, statistical and political*, vol. 1 (Dublin, 1812), p. 364.
3 Ó Danachair, 'Flail and other threshing methods', p. 9.
4 Ibid., 1; E.E. Evans, *Irish folkways* (London, 1957), p. 213.
5 Caiomhín Ó Danachair, 'The flail in Ireland', *Ethnologia Europaea*, 4 (Arnhem, 1971).
6 Wakefield, *Account of Ireland, statistical and political*, vol. 1, p. 364; J. Binns, *Miseries and beauties of Ireland*, vol. 1 (London, 1837), pp 265–6; Ordnance Survey Memoirs, box 19,7 (1835), Ó Danachair, 'Flail and other threshing methods', 9.
7 Evans, *Irish folkways*, p. 215
8 Arthur Young, *A tour in Ireland* (Dublin, 1780) vol. 1, pp 218 and 231; vol. 2, 'General observations', p. 144.
9 Ó Danachair, 'Flail and other threshing methods', p. 9.
10 Wakefield, *Account of Ireland, statistical and political*, vol. 1, p. 364.
11 Martin Doyle, *A cyclopaedia of practical husbandry*, rev. ed W. Rham (London, 1844), p. 553.
12 Michael Partridge, *Farm tools through the ages* (Reading, 1973), p. 162.
13 Alan Gailey, 'Introduction and spread of the horse-powered threshing machine to Ulster's farms in the nineteenth century: some aspects', *Ulster Folklife*, 30 (Holywood, 1984), 40; John Dubordieu, *Statistical survey of the county of Down* (Dublin, 1802), pp 53–4.

14 Joseph Archer, *Statistical survey of the county of Dublin* (Dublin, 1801), p. 44; Gailey, 'Introduction and spread of the horse-powered threshing machine', 40; D.A. Beaufort, 'Materials for the Dublin Society agricultural survey of County Louth', *Journal of the County Louth Archaeological and Historical Society*, 18:1 (Dundalk, 1973), 124.

15 Gailey, 'Introduction and spread of the horse-powered threshing machine', 40.

16 *Munster Farmer's Magazine*, 3 (Cork, 1814), 34.

17 Ordnance Survey Memoirs, box 19: 8,7, 13 (1835).

18 John Sproule, *A treatise on agriculture* (Dublin, 1939), p. 78; G.E. Fussell, *The farmer's tools* (London, 1952), p. 162; Doyle, *Cyclopaedia of practical husbandry*, p. 553; *Purdon's practical farmer* (Dublin, 1863), p. 296.

19 *Northern Whig*, 7 May 1842.

20 Pierce Ltd, *The story of Pierce* (Wexford, 197?), p. 1.

21 Gailey, 'Introduction and spread of the horse-powered threshing machine', 50.

22 *Farmer's Gazette*, 8 (Dublin, 1849), 63.

23 Fussell, *The farmer's tools*, p. 166; *Farmer's Gazette*, 8 (Dublin, 1849), 63 and 81.

24 Sproule, *Treatise on agriculture*, p. 78.

25 *Munster Farmer's Magazine*, 2 (Cork, 1813), 356.

26 *Farmer's Gazette*, 8 (Dublin, 1849), 463 and 474; Congested Districts Board for Ireland, *Twelfth Annual Report* (HMSO: Dublin, 1903), p. 13.

27 Fussell, *The farmer's tools*, pp 164–5; John Sproule (ed.), *The Irish Industrial Exhibition in 1853: a detailed catalogue of its contents* (Dublin, 1854), pp 21 and 219.

28 *Purdon's practical farmer*, p. 298; Timothy P. O'Neill, *Life and tradition in rural Ireland* (London, 1977), plate 99.

29 Sproule, *Irish Industrial Exhibition in 1853*, p. 224.

30 *Purdon's practical farmer*, p. 299

31 Gailey, 'Introduction and spread of the horse-powered threshing machine', 80.

32 *Purdon's practical farmer*, p. 300.

33 *Farmer's Gazette*, 8 (Dublin, 1849), 474 and 105.

34 Joseph Archer, *Statistical survey of the county of Dublin* (Dublin, 1801), p. 44.

35 *Munster Farmer's Magazine*, 2 (Cork, 1813), 166; *Farmer's Gazette*, 7 (Dublin, 1848), 30.

36 Sproule, *Irish Industrial Exhibition in 1853*, p. 218.

37 *Farmer's Gazette*, 9 (Dublin, 1849), 342.

38 Jonathan Bell, *Ulster farming families* (Belfast, 2005), p. 71.

39 Young, *Tour in Ireland*, vol. 1, pp 36 and 74; W. Tighe, *Statistical observations relative to the county of Kilkenny* (Dublin 1802), pp 215–16.

40 Wakefield, *Account of Ireland, statistical and political*, vol. 1, p. 364.

41 Evans, *Irish folkways*, p. 213; Binns, *Miseries and beauties of Ireland*, vol. 1, pp 265–6.

42 A.T. Lucas, 'Making wooden sieves', *Journal of the Royal Society of Antiquaries of Ireland*, 81:2 (Dublin, 1951), 147–8.

43 Evans, *Irish folkways*, p. 211.

44 Mr and Mrs S.C. Hall, *Ireland: its scenery, character, etc.* vol. 1 (London), p. 83; Micheál Ó Súilleabháin, 'The bodhran, parts 1 and 2', *Treoir*, 6 (Dublin, 1974).

45 Partridge, *Farm tools through the ages*, p. 165.

46 Gailey, 'Introduction and spread of the horse-powered threshing machine', 39; Beaufort, 'Materials for the Dublin Society agricultural survey of County Louth', 129.

47 William Shaw Mason, *A statistical account, or parochial survey of Ireland*, vol. 1 (Dublin, 1814), p. 129.

48 *Purdon's practical farmer*, p. 299.

49 Partridge, *Farm tools through the ages*, p. 164.

50 *Farmer's Gazette*, 8 (Dublin, 1849), 342.

51 Binns, *Miseries and beauties of Ireland*, pp 265–6.

52 Tighe, *Statistical observations relative to the county of Kilkenny*, p. 216.

53 A.T. Lucas, 'An Fhóir: a straw rope granary', *Gwerin*, 1 (Oxford, 1957), 3.

54 Young, *Tour in Ireland*, vol. 1, p. 211.

55   Lucas, 'An Fhóir'; A.T. Lucas, 'An Fhóir: a straw rope granary: further notes', *Gwerin* 2 (Oxford, 1958).

CHAPTER THIRTEEN   *Horses and Ponies*

1   B. Smith, *The horse in Ireland* (Dublin, 1997), p. 68.
2   E. Wakefield, *An account of Ireland statistical and political* vol. 2 (London, 1812), p. 352.
3   Fergus Kelly, *Early Irish farming* (Dublin, 1997), pp 89–90, 96.
4   Smith, *The horse in Ireland*, p. 68.
5   'Kerry bog pony', Kerry Bog Village Museum, promotional leaflet, *c*.1995.
6   Kelly, *Early Irish farming*, p. 91
7   C. Coote, *Statistical survey of the county of Armagh* (Dublin, 1804), p. 291.
8   G. Sampson, *A memoir explanatory of the chart and survey of the County Londonderry* (London, 1814), p. 182.
9   J. Dubourdieu, *Statistical survey of the county of Antrim* (Dublin, 1812), p. 334.
10   Anon., 'The Irish horse breeding industry', *Ireland industrial and agricultural* (Dublin, 1902), p. 329.
11   H. Dutton, *Statistical and agricultural survey of the county of Galway* (Dublin, 1824), p. 113.
12   J.E. Ewart, 'The ponies of Connemara', *Ireland industrial and agricultural* (Dublin, 1902), pp 332–41.
13   Anon., 'The horse in Ireland', *Journal of the Department of Agriculture and Technical Instruction*, 5:1 (Dublin, 1904), 19.
14   Mervyn Watson, 'Cushendall hill ponies', *Ulster Folklife*, 26 (1980), 13.
15   Mervyn Watson, 'The role of the horse on Irish farms' in Trefor Owen (ed.), *From Corrib to Cultra* (Belfast, 2000), p. 128.
16   J. Hanly, *Mixed farming: a practical text in Irish agriculture* (Dublin, 1924).
17   Smith, *The horse in Ireland*, p. 276.
18   Anon., 'The Irish horse breeding industry', p. 327.
19   Smith, *The horse in Ireland*, p. 28.
20   Annual General Report of the Department of Agriculture and Technical Instruction, 1904–5, quoted in Colin A. Lewis, 'Recognition and development of the Irish draught horse', Mary McGrath and Joan C. Griffith (eds), *The Irish Draught Horse: a history* (Cork, 2005), p. 149.
21   'Notes and Memoranda', *Journal of the Department of Agriculture and Technical Instruction*, 8:3 (1918), 369.
22   Watson, 'The role of the horse on Irish farms', p. 130.
23   Ibid., p. 131.
24   Thomas Baldwin, *Introduction to Irish farming* (London, 1874), p. 81.
25   Ewart, 'The ponies of Connemara', 349.
26   Watson, 'Cushendall Hill Ponies', 12.
27   'Kerry Bog Pony'.
28   D. Coleman, 'The Kerry Bog Pony', *The Chase Journal*.
29   Anon., 'The Irish horse breeding industry', p. 327.
30   Smith, *The horse in Ireland,* pp 278 and 298.
31   P.J. O'Donovan, *The economic history of livestock in Ireland* (Dublin, 1940), p. 382.
32   Mervyn Watson, 'Backins, hintins, ins and outs and turnouts: a study of horse ploughing societies in Northern Ireland' (MA dissertation, Queens University Belfast, 1991).
33   A. O'Molloy, 'Power unit of the future: the Irishmen who stick by their native draught breed' *Heavy Horse World*, 12:1 (1998), 18.

CHAPTER FOURTEEN   *Cattle*

1   Sir William Wilde 'On the ancient and modern races of oxen in Ireland', *Proceedings of the Royal Irish Academy*, 7 (Dublin, 1835).

2   R. Wallace, *Farm livestock of Great Britain* (Edinburgh 1923), p. 210.
3   Leo Curran, *Kerry and Dexter cattle* (Dublin, 1990), pp 29–30.
4   D. Low, *On the domesticated animals of the British Isles* (London, 1845) p. 580.
5   Ibid.   6 Ibid.
7   Wallace, *Farm livestock of Great Britain*, p. 211.
8   Curran, *Kerry and Dexter cattle*, p. 31.
9   Ibid., pp 29–32.   10 Ibid., p. 33.   11 Ibid., p. 44.
12 Low, *On the domesticated animals of the British Isles*, p. 580.
13 Ibid., pp 85–7.
14 Curran, *Kerry and Dexter cattle*, p. 39.
15 Ibid.   16 Ibid., pp 47–9.
17 J. Wilson, 'Irish cattle (Dexters and Kerrys)' in R.P. Wright (ed.), *The standard cyclopaedia of modern agriculture & rural economy*, vol. 7 (London, 1910) p. 152.
18 Ibid., p. 153.
19 Wallace, *Farm livestock of Great Britain*, p. 214.
20 Ibid., p. 214.
21 Curran, *Kerry and Dexter cattle*, p. 61.
22 Wallace, *Farm livestock of Great Britain*, p. 212.
23 Ibid., p. 217.   24 Ibid., pp 212–13.
25 Curran, *Kerry and Dexter cattle*, p. 54.
26 Wallace, *Farm livestock of Great Britain*, p. 211.
27 Ibid., p. 211.
28 Curran, *Kerry and Dexter cattle*, pp 88–9.
29 Ibid.   30 Ibid.   31 Ibid.
32 Finbar McCormick, personal comment.
33 D. Low, *The breeds of domestic animals of the British Isles*, vol. 2 (London, 1835).
34 P.H. Fox, 'The Irish Moil', *Irish Homestead*, 5 August 1904 (Dublin), p. 276.
35 *Wallco, Farm livestock of Great Britian*, pp 86–7.
36 *Irish Homestead*, 12 April 1919 (Dublin), p. 261.
37 Robert Wallace, *Farm livestock of Great Britain* (5th ed.); Leo Curran, *Kerry and Dexter cattle* (Dublin, 1990), pp 5–6.
38 P.W. Joyce, *A social history of ancient Ireland* vol. 2 (2nd ed.) (1913), p. 278; Jonathan Bell, 'Last sheaves, ancient cattle and Protestant bibles', *Béaloideas* 53 (Dublin, 1985).
39 Irish Moiled Cattle Society Diamond Jubilee Commemorative Leaflet 1986.
40 Ibid.   41 Ibid.   42 Ibid.
43 Curran, *Kerry and Dexter cattle*, p. 14.
44 Wright, *Standard cyclopaedia of modern agriculture*, p. 119.
45 Curran, *Kerry and Dexter cattle*, pp 16–17.
46 Brendan Dunford, *Farming and the Burren* (Dublin, 2002), p. 41.
47 Ibid., p. 17.
48 Wright, *Standard cyclopaedia of modern agriculture*, p. 119.
49 Ibid., p. 121.
50 Curran, *Kerry and Dexter cattle*, p. 14.
51 Lord George Hill, *Facts From Gweedore* (Belfast, 1971), p. 25.
52 Curran, *Kerry and Dexter cattle*, p. 10.
53 Jonathan Bell, 'Cattle', *Oxford companion to Irish history* ed. S. Donnelly (Oxford, 1998).
54 Cormac Ó Gráda, *Ireland: a new economic history, 1780–1939* (Oxford, 1994), pp 32–3, and 268.
55 *Dublin Penny Journal*, 1 (1832), 143.
56 Dunford, *Farming and the Burren*, p. 39.
57 E.E. Evans, *Mourne County* (1951), p. 115.
58 Thomas Baldwin, *Introduction to practical farming* (Dublin, 1877), p. 101.
59 John Mogey, *Rural life in Northern Ireland* (London, 1947), p. 97.
60 Ó Gráda, *Ireland: a new economic history*, p. 161.

61 Ibid., p. 417.
62 Ibid.; Baldwin, *Introduction to practical farming*, pp 78–94.
63 Baldwin, *Introduction to practical farming*, pp 78–94.
64 Anon., 'The dairying industry in Ireland', *Ireland industrial and agricultural* (Dublin, 1902), p. 236.
65 Colin Rynne, 'Butter' in Brian Lalor (ed.), *The encyclopaedia of Ireland* (Dublin, 2003), p. 146.
66 Anon., 'The dairying Industry in Ireland', p. 236.
67 Ibid., p. 240.
68 R.A. Anderson, 'Agricultural co-operation in Ireland', *Ireland industrial and agricultural* (Dublin, 1902), pp 219–20.
69 Ibid., p. 221.
70 Mary E. Daly, *The first department* (Dublin, 2002), pp 47–8.
71 Patrick Bolger, *The Irish co-operative movement* (Dublin, 1977).
72 Daly, *The first department*, p. 48.   73 Ibid., p. 77.
74 Ordnance Survey memoirs Carrickfergus, p. 240.
75 Baldwin, *Introduction to practical farming*, p. 87.

CHAPTER FIFTEEN *Sheep*

1 George Cully, *Observations on livestock* (London, 1807), p. 166.
2 Arthur Young, *A tour in Ireland*, vol. 2, part 1, (1780), p. 163.
3 Hely Dutton, *Statistical survey of the county of Galway* (Dublin, 1824), pp 114–15.
4 W. Youatt, *Sheep, their breeds, management and diseases* (London, 1837), p. 358.
5 Department of Agriculture and Technical Instruction for Ireland, 'Sheep breeding in Ireland', *Ireland industrial and agricultural* (Dublin, 1902), p. 364.
6 *Stephen's book of the Farm* (5th ed.), vol. 3 (London, 1908), p. 157; Thomas Baldwin, *Introduction to Irish farming* (Dublin, 1877), p. 109.
7 J. Hanly, *Mixed farming: a practical text in Irish agriculture* (Dublin, 1924), p. 452.
8 Webpage of the Galway Sheep Breeders Society of Great Britain: www.sheep.ukf.net/Galway3.
9 *Irish Farmer's Gazette*, 6 June 1863, p. 200.
10 Young, *Tour in Ireland*, vol 2, part 1, p. 275.
11 David Low, *The domesticated animals of the British Isles* (1845), p. 74
12 Department of Agriculture and Technical Instruction, 'Sheep breeding in Ireland', p. 366.
13 H.M. Fitzpatrick, *Ireland's countryside* (Dublin, 1972), p. 45.
14 Department of Agriculture and Technical Instruction, 'Sheep breeding in Ireland', p. 366.
15 R.F. Scharff, 'Some notes on the Irish sheep', *Irish Naturalist*, July 1922, p. 74.
16 Department of Agriculture and Technical Instruction, 'Sheep breeding in Ireland', p. 366.
17 Low, *The domesticated animals of the British Isles*, p. 74.
18 Department of Agriculture and Technical Instruction, 'Sheep breeding in Ireland', p. 367.
19 Ibid., pp 366–8.
20 John Dubordieu, *Statistical survey of the county of Down* (Dublin, 1802), p. 302.
21 *Irish Farmer's Gazette*, 9 April 1864, p. 142.
22 Mervyn Watson, 'Il y a des fagots et des fagots', *Ulster Folk and Transport Museum Yearbook* (Holywood, 1981).
23 Information from Nora Cooper, Aughavilla, Warrenpoint, Co. Down (Ulster Folk and Transport Museum archives V.16.3).
24 Young, *Tour in Ireland*, vol. 2, part 1, p. 457.
25 Young, *Tour in Ireland*, vol. 1 p. 88.
26 Cully, *Observations on livestock*, p. 166.
27 Lord George Hill, *Facts from Gweedore* (Belfast, 1971), p. xv.
28 W.E. Vaughan, *Sin, sheep and Scotsmen* (Belfast, 1982).
29 Jonathan Bell, 'Sheep' in S.J. Connolly (ed.), *The Oxford companion to Irish history* (Oxford, 1998), p. 510.

30   Department of Agriculture and Technical Instruction for Ireland, 'Sheep breeding in Ireland',
       p. 364.
31   Ibid., p. 368.   32 Young, *Tour in Ireland* vol. 1, p. 90.
33   Ibid., p. 275.   34 Ibid., p. 66.
35   Thomas Baldwin, *Introduction to Irish farming* (Dublin, 1877), p. 111.
36   Young, *Tour in Ireland*, vol. 2, part 1, pp 96, 117, and 125.
37   Ibid., vol.1, p. 83.   38 Ibid., vol. 2, 'Recommendations', p. 219.
39   Ibid., vol. 1, p. 201.   40 Ibid., vol. 2, part 1, p. 146.
41   G.V. Sampson, *Statistical survey of the county of Londonderry* (Dublin, 1802), pp 221–2.
42   Jonathan Bell, *Ulster farming families* (Belfast, 2005), p. 39.
43   Young, *Tour in Ireland*, vol. 2, part 1, p. 126.
44   Ibid., p. 14.
45   Low, *The domesticated animals of the British Isles*, p. 74.
46   Iris Combe, *Herding dogs: their origins and development in Britain* (London, 1987), p. 131.
47   H.D. Richardson, *Dogs: their origin and varieties* (Dublin, 1847), pp 83–4.
48   Bell, *Ulster farming families*, p. 37.
49   Fred Coll and Jonathan Bell, 'An account of life at Machaire Gathlán, County Donegal', *Ulster
       Folklife*, 36 (Holywood, 1990).
50   John Mogey, *Rural life in Northern Ireland* (London, 1947), p. 97.
51   John Sproule, *A treatise on agriculture* (Dublin, 1839), p. 668.
52   Ibid.
53   *Ireland, industrial and agricultural*, p. 265; W.L. Micks, *An account of the Congested Districts
       Board of Ireland* (Dublin, 1925), p. 29.
54   Ibid., p. 670.
55   Young, *Tour in Ireland*, vol. 2, part 1, p. 11.   56 Ibid., *Tour in Ireland*, vol. 1, p. 275.
57   Ibid., p. 283 and vol. 2, part 1, pp 86 and 121.
58   Mairead Dunlevy, 'Woollens' in Brian Lalor (ed.), *The encyclopaediea of Ireland* (Dublin, 2003),
       pp 1152–3.
59   Regina Sexton, 'Meat' in Brian Lalor (ed.), *The encyclopaediea of Ireland* (Dublin, 2003), p. 705.
60   Baldwin, *Introduction to Irish farming*, p. 107, p. 107.
61   Ulster Folk and Transport Museum tape R90.81.
62   Robert Jennings, *Sheep, swine and poultry* (Philadelphia, 1863), pp 190–1.
63   Margaret Donnelly, 'Meat Industry' in Brian Lalor (ed.), *The encyclopaediea of Ireland* (Dublin,
       2003), p. 706.
64   Eric Donald, 'Agribusiness' in Brian Lalor (ed.), *The encyclopaediea of Ireland* (Dublin, 2003), p.11.

CHAPTER SIXTEEN   *Pigs*

1   Alexander W. Shaw, 'The bacon curing industry' in *Ireland industrial and agricultural* (Dublin,
       1902), p. 241.
2   Mervyn Watson, 'The role of the pig in food conservation and storage in traditional Irish
       farming' in J.P. Pals and L. van Wijngaarden (eds), *Seasonality* (Environmental Archaeology no. 3)
       (Swarsluis, 1994).
3   Shaw, 'The bacon curing industry', p. 242.   4 Ibid.
5   Watson, 'Role of the pig in food conservation …', 1.
6   Horatio Townsend, 'On sheep and pigs', *Munster Farmer's Magazine*, 2 (Cork, 1813), 121.
7   *Irish Farmer's Journal and Weekly Intelligence*, 7 (Dublin, 1819), 202.
8   H.D. Richardson, *Pigs; their origin and varieties* (Dublin, 1847), p. 32.
9   *British Farmer's Magazine* vol. 22 No 63 (1852), p. 221; Shaw, 'The bacon curing industry', p. 249.
10   Ibid.   11 Ibid.
12   Ibid., pp 249–50.   13 Ibid., p. 246

14   R. Wallace, *Farm livestock of Great Britain* (Edinburgh, 1923), pp 813–14.

15   John Dubordieu, *Statistical survey of the county of Antrim* (Dublin, 1812), pp 340–4.

16   John Dubordieu, *Statistical survey of the county of Down* (Dublin, 1802), p. 207.

17   Shaw, 'The bacon curing industry', p. 249.

18   S. Hulme, *Book of the pig* (Surrey, 1979), p. 40.

19   Wallace, *Farm livestock of Great Britain*, p. 398.

20   W.E. Coey, 'The pig in Ireland' in A.E. Muskett (ed.), *A.A. McGuckian: a memorial volume* (Belfast, 1956), p. 61.

21   Ibid.

22   Government of Northern Ireland, Ministry of Agriculture, 'The Bacon Curing Industry', *Monthly Agricultural Report,* 29:2 (Belfast, 1954), 40.

23   Ulster Folk and Transport Museum tape R.86.248

24   Mervyn Watson, 'Standardisation of pig production: the case of the Large White Ulster pig', *Ulster Folklife*, 34 (Holywood, 1988).

25   Ibid., p. 6.

26   S. Hoctor, *The department's story* (Dublin, 1971), p. 151.

27   Government of Northern Ireland, Ministry of Agriculture, *Monthly Report*, 9:9 (HMSO: Belfast, 1934), p. 6

28   Ibid., p. 34.

29   L. Symons, 'The agricultural industry, 1921–62' in L. Symons (ed.), *Land use in Northern Ireland* (London, 1965), p. 43.

30   Government of Northern Ireland, Ministry of Agriculture, *Monthly Report*, 9:11 (HMSO: Belfast, 1935), p. 11.

31   Patrick MacGill, *Children of the dead end* (London, 1914), pp 35–6.

32   E.J. Sheehy, 'Pig feeding' in A.E. Muskett (ed.), *A.A. McGuckian: a memorial volume* (Belfast, 1956), p. 88.

33   Dubordieu, *Statistical survey of the county of Antrim*, pp 342–3.

34   Richardson, *Pigs; their origin and varieties*, p. 57.

35   Ibid., p. 63.

36   Information from Mr Malachy McSparran, Cloney, Cushendun.

37   Jonathan Bell, *Ulster farming families* (Belfast, 2005), pp 92–3.

38   Shaw, 'The bacon curing industry', pp 241–4

39   Watson, 'Role of the pig in food conservation …', 4.

40   Shaw, 'The bacon curing industry', p. 254.

41   Ibid.

42   T. Gaffikin, *Belfast fifty years ago* (Belfast, 1894), p. 29.

43   P.R. Wright (ed.), *The standard cyclopedia of modern agriculture and rural economy*, vol. 12 (London, 1910), p. 32.

44   Government of Northern Ireland, *The marketing of Northern Ireland produce* (HMSO: Belfast, 1932), p. 105.

45   Coey, 'The pig in Ireland', p. 61.

46   Watson, 'Role of the pig in food conservation …', 7.

47   Richardson, *Pigs; their origin and varieties*, p. 97.

48   Watson, 'Role of the pig in food conservation …', 10–12.

49   Government of Northern Ireland , Ministry of Agriculture, *Monthly Report*, 29:2 (HMSO: Belfast, 1954), 43.

50   Shaw, 'The bacon curing industry', p. 244

51   Ibid., p. 254.   52 Ibid., p. 245.

53   Wikipedia.org/wiki/Mitchelstown; Eric Donald, 'Agribusiness' in Brian Lalor (ed.), *The encyclopaedia of Ireland* (Dublin, 2003), p. 11.

CHAPTER SEVENTEEN *Poultry*

1   Arthur Young, *A tour in Ireland* vol. 2, part 1, (Dublin, 1780), p. 39.
2   Martin Doyle, *A cyclopaedia of practical husbandry* (rev. ed.) W. Rham (London, 1844), p. 472.
3   Thomas Baldwin, *Introduction to practical farming* (Dublin, 1877), p. 117.
4   www.wildwnc.org/af/rhodeislandred.html.
5   John Sproule, *A treatise on agriculture* (Dublin, 1839), p. 110.
6   *Purdon's practical farmer* (Dublin, 1863), pp 466–7.
7   Baldwin, *Introduction to practical farming*, p. 171.
8   Doyle, *A cyclopaedia of practical husbandry*, p. 472.
9   Ibid., p. 473.   10 Ibid., pp 475–6.
11 Ibid., p. 473.
12 Ibid., pp 474–5.   13 Ibid., pp 476–7.
14 Ibid., pp 479–82.   15 Ibid., pp 483–4.
16 Young, *Tour in Ireland*, vol. 1 pp 296–7 and 366.
17 Doyle, *A cyclopaedia of practical husbandry*, p. 486.
18 Ibid., p. 486.
19 *Purdon's practical farmer*, p. 467.
20 Baldwin, *Introduction to practical farming*, p. 171.
21 Patrick Bolger, *The Irish co-operative movement* (Dublin, 1977), p. 280.
22 Catherine Webb, quoted in Bolger, *The Irish co-operative movement*, p. 278.
23 Ibid., pp 279–86.
24 Bolger, *The Irish co-operative movemen*, p. 283.
25 Joanna Bourke, 'Women and poultry in Ireland, 1891–1914', *Irish Historical Studies*, 25:99 (May 1987), 301–6.
26 Mary E. Daly, *The first department* (Dublin, 2002), p. 59.
27 Bolger, *The Irish co-operative movement*, p. 286.
28 Bourke, 'Women and poultry in Ireland', 301–6.
29 Rosemary Harris, *Prejudice and tolerance in Ulster* (Manchester, 1972), p. 53.
30 Jonathan Bell, *Ulster farming families* (Belfast, 2005), pp 54–5.
31 Ibid., p. 33.
32 Bolger, *The Irish co-operative movement*, p. 287.

CHAPTER EIGHTEEN *The Theory and Practice of Improvement*

1   Mary E. Daly, *The first department: a history of the Department of Agriculture* (Dublin, 2002), p. 52.
2   Ibid., p. 102.
3   Ibid., pp 138–9.   4 Ibid., pp 159–60.
5   Jonathan Bell, *Ulster farming families* (Belfast, 2005), pp 15–17.
6   Daly, *The first department*, p. 56.
7   www.ssbtractor.com/features/Ford_tractors.html.
8   www.fergusonsociety.co.uk.   9 Bell, *Ulster farming families*, p. 18.
10 Peter Solar, 'Agricultural productivity and economic development in Ireland and Scotland in the early nineteenth century', in T.M. Devine and D. Dickson (eds), *Ireland and Scotland, 1600–1850* (Edinburgh, 1983), p. 76.
11 W. Tighe, *Statistical observations relative to the county of Kilkenny* (Dublin, 1802), p. 178.
12 Charles W. Hamilton, 'Abstract of answers to the Royal Dublin Society's agricultural queries of 1859', *Royal Dublin Society Journal*, 3 (Dublin, 1860), 24–5.
13 *Irish Farmer's Gazette*, 22 (Dublin, 1863), 483; *Irish Farmer's Gazette*, 23 (Dublin, 1864), 135.
14 *Purdon's practical farmer* (Dublin, 1863), pp 198–9.
15 P.R. Wright (ed.), *The standard cyclopaedia of modern agriculture and rural economy*, vol. II (London, 1908), p. 112.

16 Edward Burroughs, *The Irish farmer's calendar* (Dublin, 1835), p. iii.

17 James McParlan, *Statistical survey of the county of Mayo* (Dublin, 1802), p. 37.

18 Arthur Young, *Tour in Ireland*, vol. 2, part 1, (Dublin, 1780), pp 81–2.

19 Tighe, *Statistical observations relative to the county of Kilkenny*, p. 410

20 W.L. Micks, *An account of the Congested Districts Board for Ireland from 1891 to 1923* (Dublin, 1925), p. 243.

21 Hely Dutton, *Statistical survey of the county of Clare* (Dublin, 1808), p. 37.

22 Anon., 'On Agricultural Schools', *Irish Farmer's and Gardener's Magazine*, vol. 2 (Dublin, 1835), pp 402–3.

23 Young, *Tour in Ireland*, vol. 2, part 1, p. 138.

24 Ibid., pp 138–9.

25 R. Townsend, *Statistical survey of the county of Cork* (Dublin, 1810), p. 544.

26 John Sproule, *A treatise on agriculture suited to the soil and climate of Ireland* (Dublin, 1839), pp 171–2.

27 A.T. Lucas, 'Irish ploughing practices, part two', *Tools and Tillage*, 2:2 (Copenhagen, 1973), 72–9.

28 *Purdon's practical farmer*, pp 1–2.

29 David Edgerton, *The shock of the old: technology and global history since 1900* (London, 2007).

30 Joan Thirsk, 'Agricultural innovations and their diffusion', in Joan Thirsk (ed.), *The agrarian history of England and Wales*, 5:2 (Cambridge, 1985), p. 587.

31 Bell, *Ulster farming families*, pp 18–19.

32 Solar, Peter, op. cit., pp 80–1.

33 T.M. Devine and David Dickson, 'A review of the symposium', in T.M. Devine and D. Dickson (eds), *Ireland and Scotland, 1600–1800* (Edinburgh, 1983), p. 269.

Bibliography

Aalen, F.H.A., 'Public housing in Ireland, 1880–1921', *Planning Perspectives*, 2 (London, 1987).

——, 'Buildings' in Aalen et al. (eds), *Atlas of the Irish rural landscape* (Cork, 1997).

Adams, G.B., 'Work and words for haymaking', *Ulster Folklife*, 12 (Holywood, Co. Down, 1966).

Agricultural classbook (Dublin, 1856).

Anderson, James, 'Rundale, rural economy and agrarian revolution' in William Nolan, Liam Roynayne & Mairead Dunlevy (eds), *Donegal; history and society* (Dublin, 1995).

Anderson, R.A., 'Agricultural co-operation in Ireland', *Ireland industrial and agricultural* (Dublin, 1902).

Andrews, J.H., 'Limits of agricultural settlement in pre-Famine Ireland' in L.M. Cullen & F. Furet (eds), *Ireland and France, 17th–20th centuries* (Paris, 1980).

Anon., 'The dairying industry in Ireland', *Ireland industrial and agricultural* (Dublin, 1902).

Anon., 'The horse in Ireland', *Journal of the Department of Agriculture and Technical Instruction*, 5:1 (Dublin, 1904).

Anon., 'Kerry Bog Pony', Kerry Bog Village Museum, promotional leaflet, *c.*1995.

Anon., *The story of Pierce* (Wexford, 197?)

Archer, Joseph, *Statistical survey of the county Dublin* (Dublin, 1801).

Arensberg, Conrad, *The Irish countryman* (Gloucester, MA, 1959).

The Ark (Belfast, 1910).

Baker, John Wynn *Experiments in agriculture, made under the direction of the Right Honourable Dublin Society in the Year 1767* (Dublin, 1769).

——, *A short description and list, with the prices of the instruments of husbandry, made in the factory at Laughlinstown, near Cellbridge, in the county of Kildare* (Dublin, 1767–69).

Baldwin, Thomas, *Handy book of small farm management* (Dublin, 1870).

——, *Introduction to Irish farming* (London, 1874).

——, *Introduction to practical farming* (Dublin, 1877).

——, *Introduction to practical farming* 23rd ed. (Dublin, 1893).

Beatson, Robert, 'On cottages', *Irish Agricultural Magazine* (Dublin, 1793).

Beaufort, D.A., 'Materials for the Dublin Society agricultural survey of County Louth', *Journal of the County Louth Archaeological and Historical Society*, 18:1 (Dundalk, 1973).

Bede, *A history of the English church and people* trans. Leo Sherley-Price (rev. ed.) R.E. Latham (London, 1983).

Belfast Commercial Chronicle, 1 October 1806.

Belfast Newsletter, 3 August 1750.

Bell, Jonathan & Mervyn Watson, *Irish farming: implements and techniques, 1750–1900* (Edinburgh, 1986).

—— & Mervyn Watson, *Farming in Ulster* (Belfast, 1988), p. 5.

—— & Mervyn Watson, 'Cultivation ridges in County Longford', *Archaeology Ireland,* 19:4 (Dublin, 2005).

—— & Mervyn Watson, 'A horse plough and harrows from County Leitrim' in Marion Meek (ed.), *The modern traveller to our past* (Belfast, 2006).

—— & Mervyn Watson, 'Clachan project: a report for the Glens of Antrim Historical Society' (unpublished), 2006.

Bell, Jonathan, 'Straw harness in Ireland', *Ulster Folklife,* 24 (1978).

——, 'Wooden ploughs from the Mountains of Mourne, Ireland', *Tools and Tillage,* 4:1 (Copenhagen, 1980).

——, 'Sickles, hooks and scythes in Ireland', *Folklife,* 19 (Leeds, 1981).

——, 'Harrows used in Ireland', *Tools and Tillage* 4:4 (Copenhagen, 1983).

——, 'Recent evidence for the use of Ard Ploughs in Ireland', *Sinsear* (Dublin, 1983).

——, 'The use of oxen on Irish farms since the eighteenth century', *Ulster Folklife,* 29 (Holywood, 1983).

——, 'A contribution to the study of cultivation ridges in Ireland', *Journal of the Royal Society of Antiquaries of Ireland,* 114 (Dublin, 1984).

——, 'Last sheaves, ancient cattle and Protestant Bibles', *Béaloideas,* 53 (Dublin, 1985).

——, 'Miserable hovels and substantial habitations: the housing of rural labourers in Ireland since the eighteenth century', *Folklife,* 34 (Reading, 1996).

——, *Ulster farming families* (Belfast, 2005).

Berry, Henry F., *A history of the Royal Dublin Society* (London, 1915).

Bigger, Francis Joseph, 'Labourer's cottages for Ireland', *Independent Newspaper* (Dublin, 1907).

Binchy, D.A. (ed.), *Crith Gablach* (Dublin, 1941).

Binns, J., *Miseries and beauties of Ireland,* 2 vols (London, 1837).

Board of Agriculture (England), 'Haymaking', Leaflet no. 85, Department of Agriculture and Technical Instruction for Ireland, *Journal,* 3 (Dublin: HMSO, 1903).

Bolger, Patrick, *The Irish co-operative movement* (Dublin, 1977).

Bourke, A., *The visitation of God?: the potato and the Great Irish Famine*, ed. Jacqueline Hill and Cormac Ó Gráda (Dublin, 1993).

Bourke, Joanna, 'Women and poultry in Ireland, 1891–1914', *Irish Historical Studies* 25:99 (May 1987).

Boyle, John W., 'A marginal figure: the Irish rural labourer' in S. Clarke and J.S. Donnelly (eds), *Irish peasants: violence and political unrest, 1780–1914* (Manchester, 1983).

British Farmer's Magazine vol. 22:63 (1852).

Buchanan, R., 'Calendar customs, part 1' *Ulster Folklife,* 8 (Holywood, 1962).

Buchanan, R., 'Field systems in Ireland' in (eds), A.R.H. Baker and R.A. Butlin, *Studies in the field systems of the British Isles* (Cambridge, 1973).

Burroughs, E., *The Irish farmer's calendar* (Dublin, 1835).

Cassell's Household Guide, vol. 4 (Cassells: London, 1880).

Caulfield, Seamas, 'Neolithic fields: the Irish evidence' H. Bowen and P. Fowler (eds), *British Archaeological Report*, 48 (Oxford, 1978).

Clark, A.L., 'Historical sketch of the flax-growing industry' Department of Agriculture and Technical Instruction for Ireland, *Journal*, vol. 3 (HMSO: Dublin, 1903).

——, 'Flax seed for sowing purposes', Department of Agriculture and Technical Instruction for Ireland, *Journal,* vol. 4 (HMSO: Dublin, 1904).

Coey, W.E., 'The pig in Ireland' in A.E. Muskett (ed.), *A.A. McGuckian: a memorial volume* (Belfast, 1956).

Coleman, D., 'The Kerry Bog Pony', *The Chase Journal.*

Coll, Fred and Jonathan Bell, 'An account of life in Machaire Gathlán, County Donegal', *Ulster Folklife,* 36 (Holywood, 1990).

Collins, E.J.T., *Sickle to combine* (Reading, 1969).

Collins, James, *Quickening the earth: soil minding and mending in Ireland* (Dublin, 2008).

Combe, Iris, *Herding dogs: their origins and development in Britain* (London, 1987).

Congested Districts Board for Ireland, *Twelfth Annual Report* (HMSO: Dublin, 1903).

Connell, K.H., 'The colonisation of waste land in Ireland, 1780–1845', *Economic History Review* 3:1 (London, 1950).

Coote, C., *Statistical survey of the county of Cavan* (Dublin, 1802).

——, *Statistical survey of the county of Armagh* (Dublin, 1804).

Cotter, Rev. G.S., 'Advice to the Farmers of the county of Cork, on the culture of wheat', *Munster Farmer's Magazine,* 1 (Cork, 1812).

County Agricultural Instructors, *Agriculture in Ireland* (Dublin, 1907).

Craig, E.T., *An Irish commune: the history of Ralahine* (Dublin, 1919).

Crawford, W.H., 'Economy and society in eighteenth-century Ulster', PhD thesis (Queen's University Belfast, 1982).

Cullen, L.M., *Life in Ireland* (London, 1968).

——, *An economic history of Ireland since 1600* (London, 1972).

Cully, George, *Observations on livestock* (London, 1807).

Curran, Patrick Leonard, *Kerry and Dexter cattle and other ancient Irish breeds* (Dublin, 1900).

Curran, Simon, 'The Society's role in agriculture since 1800' in J. Meenan and D. Clarke (eds), *The Royal Dublin Society, 1731–1981* (Dublin, 1981).

Daly, Mary E., *The first department: a history of the Department of Agriculture* (Dublin, 2002).

Davies, C., 'Kilns for flax-drying and lime-burning', *Ulster Journal of Archaeology*, 3rd series, 1 (Belfast, 1938).

Davis, Francis, 'Low and clean' in *Earlier and later leaves* (Belfast, 1878).

De Latocnaye, *A Frenchman's walk through Ireland, 1796–7* trans. J. Stevenson (Belfast, 1917).

Dennehy, Mary, 'Agricultural education in the nineteenth century', *Retrospect: Journal of the Irish History Students' Association,* 5 (Dublin, 1982).

Department of Agriculture and Technical Instruction for Ireland, 'Flax experiments, 1901', *Journal,* 2 (HMSO: Dublin, 1902).

Department of Agriculture and Technical Instruction for Ireland, 'Prevention of potato blight', *Journal,* 2 (HMSO: Dublin, 1902).

Department of Agriculture and Technical Instruction for Ireland, 'Statistical survey of Irish agriculture', *Ireland, industrial and agricultural* (Dublin, 1902).

Department of Agriculture and Technical Instruction for Ireland, 'Haymaking', Leaflet no. 46, *Journal* vol. 4 (HMSO: Dublin, 1904).

Department of Agriculture and Technical Instruction for Ireland *Twelfth Annual Report, 1911–1912* (HMSO: Dublin, 1913).

Devine, T.M. & David Dickson, 'A review of the symposium' in T.M. Devine & David Dickson (eds), *Ireland and Scotland, 1600–1800* (Edinburgh, 1983).

[Devon Commission], *Digest of evidence taken before Her Majesty's Commissioners of Inquiry into the state of the law and practice in respect to the occupation of land in Ireland,* 2 parts (Dublin, 1847).

Dineen, Patrick S., *Foclóir Gaedhilge agus Bearla* (Dublin, 1927).

Donald, Eric, 'Agribusiness' in Brian Lalor (ed.), *The encyclopaedia of Ireland* (Dublin, 2003).

Donnelly, Margaret 'Meat Industry' in Brian Lalor (ed.), *The encyclopaedia of Ireland* (Dublin, 2003).

Downey, Liam, *A foresight of rural Ireland* (Dublin, 2003).

Doyle, Martin, *Hints originally intended for the small farmers of the county of Wexford* (Dublin, 1830).

——, *Hints to smallholders on planting and on cattle etc.* (Dublin, 1830).

——, 'On agricultural schools', *Irish Farmer's and Gardener's Magazine,* 1 (Dublin, 1834).

——, *A cyclopaedia of practical husbandry* (rev. ed.) W. Rham (London, 1844).

Dublin Penny Journal, 1–2 (Dublin, 1833–4).

Dubois, Edward, *My pocket book; or, Hints for 'A ryghte merrie and conceitede' tour* (London, 1808).

Dubordieu, John, *Statistical survey of the county of Down* (Dublin, 1802).

——, *Statistical survey of the county of Antrim* (Dublin, 1812).

Duignan, Michael 'Irish agriculture in early historic times', *Journal of the Society of Antiquaries of Ireland,* 74 (Dublin, 1944).

Dunford, Brendan, *Farming and the Burren* (Dublin, 2002).

The Dungannon News, 1 May 1902.

Dunlevy, Mairead, 'Woollens' in Brian Lalor (ed.), *The encyclopaediea of Ireland* (Dublin, 2003).

Dutton, Hely, *Observations on Mr Archer's Statistical survey of the county of Dublin* (Dublin, 1802).

——, *Statistical survey of the county of Clare* (Dublin, 1808).

——, *A statistical and agricultural survey of the county of Galway* (Dublin, 1824).

Edgerton, David, *The shock of the old: technology and global history since 1900* (London, 2007).

Evans, E.E., *Irish heritage* (Dundalk, 1949).

——, *Irish folkways* (London, 1957).

——, *Mourne Country* (Dundalk, 1951).

Ewart, J.E., 'The ponies of Connemara', *Ireland industrial and agricultural* (Dublin, 1902).

'F', 'Observations on the agriculture of the North of Ireland', *Irish Farmer's and Gardener's Magazine,* 1 (Dublin, 1834).

The farmer's friend (London, 1847).

The Farmer's Gazette and Journal of Practical Horticulture, 7–9 (Dublin, 1847–9).

Feehan, John, *Farming in Ireland: history, heritage and environment* (Dublin, 2003), p. 120.

Fenton, A., *Scottish country life* (Edinburgh, 1976).

Fitzpatrick David, 'The disappearance of the Irish agricultural labourer', *Irish Economic and Social History,* 7 (Dublin, 1980).

Fitzpatrick, H.M., *Ireland's countryside* (Dublin, 1972).

Fussell, G.E., *The farmer's tools, 1500–1900* (London, 1952).

Gaffikin, T., *Belfast fifty years ago* (Belfast, 1894).

Gailey, Alan, 'Irish iron-shod wooden spades', *Ulster Journal of Archaeology,* 31 (Belfast, 1968).

——, 'The typology of the Irish spade' in Alan Gailey and A. Fenton (eds), *The spade in Northern and Atlantic Europe* (Belfast, 1970).

——, 'Spade tillage in south-west Ulster and north Connaught', *Tools and Tillage,* 1:4 (Copenhagen, 1971).

——, *Spade making in Ireland* (Holywood, 1982).

——, 'Introduction and spread of the horse-powered threshing machine to Ulster's farms in the nineteenth century: some aspects', *Ulster Folklife,* 30 (Holywood, 1984).

Gailey, Alan, *Rural houses in the North of Ireland* (Edinburgh, 1984).

Glassie, Henry, *Passing the time in Ballymenone* (Philadelphia, 1982).

Government of Northern Ireland, Ministry of Agriculture, 'The bacon curing industry', *Monthly Agricultural Report,* 29:2 (Belfast, 1954).

Greig, W., *General report on the Gosford Estates in County Armagh* introd. F.M.L. Thompson and D. Tierney (HMSO: Belfast, 1976).

Grigg, David, *The dynamics of agricultural change* (London, 1982).

Hale, Thomas, *A compleat body of husbandry,* 4 vols (Dublin, 1757).

Hall, Mr & Mrs. S., *Ireland: its scenery, character etc.* vol. 1 (London, c.1850).

Hamilton, Charles W., 'Abstract of answers to the Royal Dublin Society's agricultural queries of 1859', *Royal Dublin Society Journal,* 3 (Dublin, 1860).

Hanly, J., *Mixed farming: a practical text in Irish agriculture* (Dublin, 1924).

Hannan, Robbie & Jonathan Bell, 'The *Bothóg*: a seasonal dwelling from County Donegal' in Trefor M. Owen (ed.), *From Corrib to Cultra* (Belfast, 2000).

Harkin, Maura & Sheila McCarroll, *Carndonagh* (Dublin, 1984).

Harris, Rosemary, *Prejudice and tolerance in Ulster* (Manchester, 1972).

Hill, George, *Facts from Gweedore* introd. E.E. Evans (Belfast, 1971).

Hoctor, D., *The department's story* (Dublin, 1971).

Howatson, W., 'Grain harvesting and harvesters' in T.M. Devine (ed.), *Farm servants and labourers in Lowland Scotland* (Edinburgh, 1984).

Hulme, S., *Book of the pig* (Surrey, 1979).

Hyndman, C., *A new method of raising flax* (Belfast, 1774).

Impartial Reporter, 3 August 1967.

Instructions to owners applying for loans for farm buildings (Dublin, 1856).

The Irish Farmer's and Gardener's Magazine, 1 (Dublin, 1834).

The Irish Farmer's Gazette (Dublin, 1863–4).

Irish Farmer's Journal and Weekly Intelligence, 7 (Dublin, 1819)

Irish National Schools, *Agricultural Classbook* (Dublin, 1868).

Irish National Schools, *Introduction to practical farming,* 2 vols. (Dublin, 1898).

Jefferies, Henry A., 'Raleigh, Sir Walter' in Brian Lalor (ed.), *The encyclopaedia of Ireland* (Dublin, 2003).

Jennings, Robert, *Sheep, swine and poultry* (Philadelphia, 1863).

Joyce, P.W., *A social history of ancient Ireland,* 2 vols (2 ed.) (Dublin, 1913).

Kelly, Fergus, *Early Irish farming* (Dublin, 1997).

Lambert, Joseph, *Observations on the rural affairs of Ireland* (Dublin, 1829).

——, *Agricultural suggestions to the proprietors and peasantry of Ireland* (Dublin, 1845).

Lane, Padraig, *Ireland* (London, 1974).

Lewis, Colin A., 'Recognition and development of the Irish Draught Horse' in Mary McGrath & Joan C. Griffith (eds), *The Irish Draught Horse: a history* (Cork, 2005)

Library of Useful Knowledge, *British husbandry,* 2 vols (London, 1834).

Loudon, J.C., *An encyclopaedia of agriculture* (London, 1831).

Low, David., *Elements of practical agriculture* (Edinburgh, 1834).

——, *The domesticated animals of the British Isles* (Edinburgh, 1845).

Lucas, A.T., 'Making wooden sieves', *Journal of the Royal Society of Antiquaries of Ireland,* 81:2 (Dublin, 1951).

——, 'An Fhóir: a straw rope granary', *Gwerin,* 1 (Oxford, 1957).

——, 'Paring and burning in Ireland; a preliminary survey', in Alan Gailey & A. Fenton (eds), *The spade in northern and Atlantic Europe* in (Belfast, 1970).

——, 'The "Gowl-Gob", an extinct spade type from County Mayo, Ireland', *Tools and Tillage,* 3:4 (Copenhagen, 1979).

——, 'Irish ploughing practices', 4 parts, *Tools and Tillage,* vol. 2: 1–4 (Copenhagen, 1972–5).

McCourt, Desmond, 'The rundale system in Donegal, its distribution and decline', *Donegal Annual,* 3 (1955).

——, 'Infield and outfield in Ireland', *Economic History Review,* 7 (1954–5).

McEvoy, John, *Statistical survey of the county of Tyrone* (Dublin, 1802).

McGahern, John, *The dark* (London, 1965).

MacGill, Patrick, *Children of the dead end* (London, 1914).

McGrath, Mary & Joan C. Griffith, (eds), *The Irish Draught Horse: a history* (Cork, 2005).

MacNeill, M., *The Festival of Lughnasa* (London, 1962).

McParlan, James, *Statistical survey of the county of Donegal* (Dublin, 1802).

——, *Statistical survey of the county of Leitrim* (Dublin, 1802).

——, *Statistical survey of the county of Mayo* (Dublin, 1802).

McParlan, James, *Statistical survey of the county of Sligo* (Dublin, 1802).

McCully, James, *Letters by a farmer* (Belfast, 1787).

Marshall, G., 'The "Rotherham" plough', *Tools and Tillage,* 4:3 (Copenhagen, 1982).

Mason, Thomas H., *The islands of Ireland,* 2nd ed. (London, 1938).

Mason, William Shaw, *A statistical account, or parochial survey of Ireland,* 3 vols (Dublin, 1814).

Meenan, J. and D. Clare, 'The Royal Dublin Society' in J. Meenan and D. Clarke (eds), *The Royal Dublin Society, 1731–1981* (Dublin, 1981).

Micks, W.L., *An account of the … Congested Districts Board for Ireland from 1891 to 1923* (Dublin, 1925).

Mitchell, Frank, *The Irish landscape* (London, 1976).

Mogey, John M., *Rural life in Northern Ireland* (London, 1947).

Mokyr, Joel, *Why Ireland starved* (London, 1983).

Moody, T.W., *Davitt and the Irish revolution, 1846–1882* (Oxford, 1981).

Mullen, Pat, *Man of Aran* (London, 1934).

The Munster Farmer's Magazine (Cork, 1812–17).

Murphy, Edmund, 'Agricultural report', *Irish Farmer's and Gardener's Magazine,* 1 (Dublin, 1834).

——, 'Spade husbandry', *Irish Farmer's and Gardener's Magazine,* 1 (Dublin, 1834).

——, *Irish Farmer's and Gardener's Register,* 1 (Dublin, 1863).

Murphy, Joey, 'The mower from Moygannon', *Ulster Folklife,* 12 (Holywood, 1966).

Nicholson, Asenath, 'The bible in Ireland'. Quoted in Frank O'Connor (ed.), *A book of Ireland* (Glasgow, 1959).

North West of Ireland Society Magazine, 1 (Derry, 1823).

Northern Whig, 7 May 1842.

Ó Danachair, Caoimhín, 'The flail and other threshing methods', *Journal of the Cork Historical and Archaeological Society,* 60:19 (Cork, 1955).

Ó Danachair, Caoimhín, 'The spade in Ireland', *Béaloideas,* 31 (Dublin, 1963).

Ó Danachair, Caoimhín, 'The use of the spade in Ireland' in Alan Gailey and A. Fenton (eds), *The spade in Northern and Atlantic Europe* (Belfast, 1970).

——, 'The flail in Ireland', *Ethnologia Europaea,* 4 (Arnhem, 1971).

Ó Donaill, N., *Foclóir Gaeilge-Bearla* (Dublin, 1977).

O'Donovan, P.J., *The economic history of livestock in Ireland* (Dublin, 1940).

O'Dowd, Anne, *Meitheal: a study of co-operative labour in rural Ireland* (Dublin, 1981).

Ó Gráda, Cormac, 'The investment behaviour of Irish landlords, 1830–75: some preliminary findings', *Economic History Review,* 23 (1975).

——, *Ireland before and after the Famine: explorations in economic history* (Manchester, 1988).

——, *Ireland: a new economic history, 1780–1939* (Oxford, 1994).

——, 'Economy, 1700–1922' in Brian Lalor (ed.), *The encyclopaedia of Ireland* (Dublin, 2003), p. 337.

O'Loan, John, 'The manor of Cloncurry, Co. Kildare, and the feudal system of land tenure in Ireland', Department of Agriculture (Ireland), *Journal,* 58 (Dublin, 1961).

——, 'A history of early Irish farming', 3 parts, Department of Agriculture (Ireland), *Journal,* 60–62 (Dublin, 1963–5).

O'Molloy, A., 'Power unit of the future: the Irishmen who stick by their native Draught breed', *Heavy Horse World,* 12:1 (1998).

O'Neill, Kevin, *Family and farm in pre-Famine Ireland* (Madison, WI, 1984).

O'Neill, Thomas P., 'The scientific investigation of the failure of the potato crop in Ireland, 1845–6', *Irish Historical Studies,* 5:18 (Dublin, 1946).

O'Neill, Timothy P., *Life and tradition in rural Ireland* (London, 1977).

O'Rahilly, T.F., 'Etymological notes, 2', *Scottish Gaelic Studies,* 2 (Edinburgh, 1927).

Ordnance survey memoirs Mss, Royal Irish Academy, Dublin.

Ó Súilleabháin, Amhlaoibh, *Cinnlae Amhlaoibh Uí Shúilleabháin,* vol. 3 trans. M. McGrath (Dublin, 1930).

Ó Súilleabháin, Micheál, 'The bodhran, parts 1 and 2', *Treoir,* 6 (Dublin, 1974).

The Oxford Companion to Irish History ed. S.J. Connolly (Oxford, 1998).

'Parliament Chloinne Tomáis' trans. O.J. Bergin, *Gadelica* (Dublin, 1912).

Parry, M.L., 'A typology of cultivation ridges in southern Scotland', *Tools and Tillage,* 3:1 (Copenhagen, 1976).

Partridge, Michael, *Farm tools through the ages* (Reading, 1973).

Patents for Inventions: Abridgments of Specifications, Class 6, Agricultural Appliances … AD 1855–6 (HMSO: London, 1905).

Perkins, J.A., 'Harvest technology and labour supply in Lincolnshire and the East Riding of Yorkshire, part 2', *Tools and Tillage,* 3:2 (Copenhagen, 1977).

Plymseley, Rev., 'On cottages' in Arthur Young (ed.), *The annals of agriculture* (London, 1804).

Proudfoot, Lindsay, 'Landlords' in S.J. Connolly (ed.), *The Oxford companion to Irish history* (Oxford, 1998), p. 297.

Purdon's practical farmer (Dublin, 1863).

Rawson, T.J., *Statistical survey of the county of Kildare* (Dublin, 1807).

Regulations for gathering seaweed, Ardglass, Co. Down. Public Records Office of Northern Ireland. Document T1009/160.

Richardson, H.D., *Pigs; their origin and varieties* (Dublin, 1847).

Roberts, Michael, 'Sickles and scythes: women's work and men's work at harvest time', *History Workshop,* 7 (Southampton, 1979).

Royal Commission on Congestion in Ireland, Second Appendix, *Digest of evidence* (HMSO: Dublin, 1908).

Russell, George (AE) (ed.), *The Irish Homestead* (Dublin, 1917).

Rynne, Colin, 'Butter' in Brian Lalor (ed.), *The encyclopaedia of Ireland* (Dublin, 2003).

Salaman, R.N., 'The influence of the potato on the course of Irish history' (Tenth Finlay Memorial Lecture) (Dublin, 1944).

——, *The history and social influence of the potato* (Cambridge, 1949).

Sampson, G.V., *A memoir, explanatory of the chart and survey of the county of London-derry, Ireland* (London, 1814).

Scharff, R.F., 'Some notes on Irish sheep', *The Irish Naturalist* (Dublin, 1922).

Scott, Paul, *A division of the spoils* (St Albans, 1975).

Sexton, Regina, 'Meat' in Brian Lalor (ed.), *The encyclopaedia of Ireland* (Dublin, 2003).

Shaw, Alexander W., 'The Irish bacon curing industry', *Ireland industrial and agricultural* (Dublin, 1902).

Sheehy, E.J., 'Pig feeding' in A.E. Muskett (ed.), *A.A. McGuckian: a memorial volume* (Belfast, 1956).

Silcock, James et al., 'The influence of rippling', Department of Agriculture and Technical Instruction for Ireland, *Journal,* 3 (HMSO: Dublin, 1903).

Smith, B., *The horse in Ireland* (Dublin, 1997).

Smith, James, *Remarks on thorough draining and deep ploughing* (Stirling, 1844).

Smith, Paddy, 'Agriculture' in Brian Lalor (ed.), *The encyclopaedia of Ireland* (Dublin, 2003).

Solar, Peter, 'Agricultural productivity and economic development in Ireland and Scotland in the early nineteenth century' in T.M. Devine and D. Dickson (eds), *Ireland and Scotland, 1600– 1850* (Edinburgh, 1983).

Sproule, John, *A treatise on agriculture* (Dublin, 1839).

—— (ed.), *The Irish Industrial Exhibition of 1853: a detailed catalogue of its contents* (Dublin, 1854).

The standard cyclopedia of modern agriculture and rural economy, 12 vols (ed. P.R. Wright) (London, 1910).

Staunton Spectator, 5 September 1839.

Stephens, Henry, *Book of the Farm* 2 vols, (3rd ed.) (Edinburgh, 1871).

Stephens' Book of the Farm, 3 vols (5th ed.) (rev. J. Macdonald) (Edinburgh, 1908).

Symons, L., 'The agricultural industry, 1921–62' in L. Symons (ed.), *Land use in Northern Ireland* (London, 1965), p. 43.

Thackeray, William Makepeace, *The Irish Sketch-book of 1842* (*Works,* vol. 18) (London, 1879).

Thirsk, Joan, 'Agricultural innovations and their diffusion' in Joan Thirsk (ed.), *The agrarian history of England and Wales,* 5:2 (Cambridge, 1985).

Thompson, R., *Statistical survey of the county of Meath* (Dublin, 1802).

Tighe, W., *Statistical observations relative to the county of Kilkenny* (Dublin, 1802).

Townsend, Horatio, *Statistical survey of the county of Cork* (Dublin, 1810).

——, 'On sheep and pigs', *Munster Farmer's Magazine,* 2 (Cork, 1813).

Trench, W.S., *Realities of Irish life* (London, 1868).

Ulster Farmer and Mechanic, 1 (Belfast, 1824–5).

Vaughan, W.E., 'Landlord and tenant relations in Ireland between the famine and the land war, 1850–78' in L.M. Cullen & T.C. Smout (eds), *Comparative aspects of Scottish and Irish economic and social history* (Edinburgh, 1978).

——, *Sin, sheep and Scotsmen* (Belfast, 1982).

Wakefield, E., *An account of Ireland, statistical and political* 2 vols (Dublin, 1812).

Wallace, M.G., 'Early potato growing', Department of Agriculture and Technical Instruction for Ireland, *Journal,* 6 (HMSO: Dublin, 1906).

Wallace, R., *Farm livestock of Great Britain* (5th ed.) (Edinburgh, 1923).

Watson, Mervyn, 'Flachters: their construction and use on an Ulster peat bog', *Ulster Folklife,* 25 (Holywood, 1979).

——, 'Cushendall Hill Ponies', *Ulster Folklife,* 26 (Holywood, 1980).

——, 'North Antrim swing ploughs: their construction and use', *Ulster Folklife,* 28 (Holywood, 1982).

——, 'Common Irish plough types and tillage practices', *Tools and Tillage,* 5:2 (Copenhagen, 1985).

——, 'Il y a des fagots et de Fagots', *Ulster Folk and Transport Museum Yearbook* (Holywood, 1981).

——, 'Standardisation of pig production: the case of the Large White Ulster pig', *Ulster Folklife,* 34 (Holywood, 1988).

——, 'Backins, hintins, ins and outs and turnouts: a study of horse ploughing societies in Northern Ireland' (MA dissertation, Queens University Belfast, 1991).

——, 'The role of the pig in food conservation and storage in traditional Irish farming' in J.P. Pals & L. van Wijngaarden (eds), *Seasonality* (Environmental Archaeology no. 3) (Swarsluis, 1994).

——, 'The role of the horse on Irish farms' in Trefor Owen (ed.), *From Corrib to Cultra* (Belfast, 2000).

——, 'Spades versus ploughs: a nineteenth century debate: part one' *Ulster Folklife,* 46 (Holywood, 2000).

——, 'Spades versus ploughs: a nineteenth century debate: part two' *Ulster Folklife,* 49 (Holywood, 2004).

Weekly Agricultural Review, 15 April 1859.

Weld, Isaac, *Statistical survey of the county of Roscommon* (Dublin, 1832).

Wexford Engineering Company, *Catalogue* (Wexford, 1950).

Wilde, W.R., 'On the ancient and modern races of oxen in Ireland', *Proceedings of the Royal Irish Academy,* 7 (Dublin, 1835).

——, *Catalogue of the antiquities of the museum of the Royal Irish Academy* (Dublin, 1857).

Wright, R.P. (ed.), *The standard cyclopaedia of modern agriculture and rural economy,*
12 vols (London, 1908).

Yager, Tom, 'What was rundale and where did it come from?', *Béaloideas,* 70 (Dublin,
2002).

Youatt, W., *Sheep: their breeds, management and diseases* (London, 1837).

Young, Arthur, *a tour in Ireland,* 2 vols (Dublin, 1780).

Young, R.M., *Historical notes of old Belfast* (Belfast, 1896).

Yule, Alexander, *Spade and manual labour, with low or cheap farming, a certain means
of removing Irish distress; applicable to England and Scotland* (Dublin, 1850).

TAPES

Interview with Mr Patrick Brogan, Hornhead, Co. Donegal.

Interview with Mr William Donnelly, Lisburn, Co. Antrim.

Interview with Mr J. Drennan, Limavady, Co. Derry.

Interview with Mr Joe Kane, Drumkeeran, Co. Fermanagh.

Interview with Richard, Eugene and Felix McConville, Dromore, Co. Down.

Interview with Mr Cormac McFadden, Roshin, Co. Donegal.

Interview with Mr John Joe McIlroy, Derrylea, Co. Monaghan.

Interview with Mr Archie Mullen, Armoy, Co. Antrim.

Interview with John O'Connor, Screen, Co. Wexford.

Interview with Mr J.A. Weir, Ballyroney, Co. Down.

Index